Mastering Your Organization's Processes

Understanding and improving your organization's business processes is vital in today's economy. Using non-technical language, this book describes the importance of these processes and the internal and external forces that shape them. It then explains the kinds of computer software available for improving and managing business processes in a flexible way. Detailed case studies illustrate that successful process management depends on attention to the human, organizational and financial factors involved, as well as the strategic implications. Finally, the book gives even-handed guidance on what to look for in Business Process Management software and discusses current technical trends.

With many clear diagrams, a glossary of terms and suggestions for further reading, the book gives the non-specialist reader a broad and informed view of business processes, free from technical imperatives. Ideal for non-technical managers, this book will also appeal to MBA and business studies students.

John O'Connell is chairman of Portrait Software plc and Clarity Commerce Solutions plc. He was founder, chairman and chief executive officer of Staffware plc until it was sold to Tibco Software, Inc. in June 2004. Staffware was acknowledged as the leading workflow and Business Process Management software company worldwide and one of Europe's top software companies.

Jon Pyke is founder and president of The Process Factory. He was chief technology officer and a main board director of Staffware plc from August 1992 until 2004. He is a world-recognized industry figure, an exceptional public speaker and a quoted company executive.

Roger Whitehead is a director of Office Futures, an independent consultancy in electronic business that he founded in 1981. The company advises organizations on using information technology to make their strategies real. It also assists with product and supplier selection, and supplier implementation.

Mastering Your Organization's Processes

A Plain Guide to Business Process Management

John O'Connell

Chairman of Portrait Software plc and Clarity Commerce Solutions plc. Formerly chairman and chief executive officer of Staffware plc.

Jon Pyke

Founder and president of The Process Factory. Formerly chief technology officer of Staffware plc.

Roger Whitehead

Director - Office Futures

CAMBRIDGE
UNIVERSITY PRESS

CAMBRIDGE UNIVERSITY PRESS

Cambridge, New York, Melbourne, Madrid, Cape Town, Singapore, São Paulo

CAMBRIDGE UNIVERSITY PRESS

The Edinburgh Building, Cambridge CB2 2RU, UK

Published in the United States of America by Cambridge University Press, New York

www.cambridge.org
Information on this title: www.cambridge.org/9780521839754

First published 2006

Printed in the United Kingdom at the University Press, Cambridge

A catalogue record for this publication is available from the British Library

ISBN-13 978-0-521-83975-4 hardback
ISBN-10 0-521-83975-0 hardback

Dedicated to the memory of Alaric Robert George Whitehead

Contents

Foreword

Richard Holway

In March 1923, in an interview with *The New York Times*, the British mountaineer George Leigh Mallory was asked why he wanted to climb Mount Everest, and replied, 'Because it's there'.

For most of the forty years that I have been involved in IT, the answer to a question like 'Why did you install that technology, that software?' was inevitably 'because IT's there'. If the dot.com madness of the late 1990s produced one lasting benefit it was, once and for all, to kill that off as a justification for the implementation of every latest IT innovation.

The business, rather than technological, case for both change and investment is now paramount. But it was not always so. Indeed terms such as business process management (BPM) and business process outsourcing (BPO) entered my lexicon as an IT analyst only in the 1990s.

I have become increasingly convinced that the way to success with complex systems is to start with clear thinking by managers and users. Any manager looking to embark on a business process change programme must first focus on how any proposed changes are likely to affect not only immediate challenges but also other processes and actions further down the line. Businesses are not assembly lines, and users have important knowledge and opinions to share.

Too many books about business processes go into elaborate detail about processes and the way computer software can help with them but forget about the business side of things. We needed a book to help to set the balance right. That's what you would expect from the authors, who have long experience of the practical side of improving organizations. It shows in the breadth of coverage they provide. They understand that business improvement is about people, about communication and about change – not just sophisticated software and powerful machines.

The aim of this book is to help any manager – no matter how non-technical – who wants to learn about the length and breadth of BPM, not just about the latest software.

Richard Holway is the Director of the research group, Ovum Holway. He has been involved in the UK IT sector for nearly forty years and is considered by many to be the UK's leading IT analyst. Richard has served as a non-executive director and, in several cases, chairman, of over a dozen private and publicly quoted UK software companies. Ovum is the largest European headquartered authority on the telecoms, software and IT services sectors.

Professor Robin Milner

To experts in business systems, computers are important players in the total performance of complex processes; the other players are human. To computer scientists, the most common thing they have to grapple with is no longer a single program running on a mainframe computer, but a network of computers serving humans in a complex interaction of tasks. So both communities confront the same phenomena, and for decades have been seeking the right concepts to express them.

I believe that business people will welcome the jargon-free descriptions, as well as the simple graded examples, that the authors use in this book. As a computer scientist, I see progress in the way that their descriptions and the concepts behind them align nicely with elementary theories of interactive systems developed since the 1960s. By 'talking through' such examples, the two communities can converge upon a shared concept of process that will serve them both, and enhance the way they work together.

Robin Milner is the Emeritus Professor of Computer Science at Cambridge and was head of the computer laboratory there from January 1996 to October 1999. He was elected Fellow of the Royal Society in 1988, and in 1991 he gained the A. M. Turing Award, the highest honour in computer science. Among his many contributions to the field, Professor Milner pioneered the pi-calculus, a means of analysing concurrent, communicating and mobile processes.

Preface

Business Process Management (BPM) is a new class of business software and, at the same time, a way of looking at organizational behaviour. This book is about using both aspects to improve and manage your organization's processes. It is intended to give people in all kinds of organization a clear understanding of the nature and importance of processes, both internal and interorganizational. It relates processes to other organizational resources and activities and gives advice on creating a process management strategy.

We try to present a balanced view. It is our firm belief that you cannot make lasting improvements to significant business processes without using computers. At the same time, we know that success comes only if you pay serious attention – and not just lip service – to the human side of organizational life.

What this book does *not* do is:

- try to sell you a method, system or product
- pretend your organization will collapse if you fail to do what we say
- talk as though the world revolves around computer systems and computer people
- argue that there is one best way of doing anything (and that we know what that way is)
- swamp you in computing jargon.

Instead, we offer you a plain, honest and balanced look at how to improve business processes.

Chapter 1 is a look at systems in general, to give you an idea of how process thinking applies in the wider world. From then on we concentrate on business processes. We describe their importance, and relate them to other processes and activities. We discuss how to change and manage processes, and how to make a case for doing so. Finally, we look at technical issues such as computing architectures and choosing a suitable product. At the end of the book are a glossary of technical terms and a list of suggested reading. Throughout, there are detailed case studies and other user testimony and examples.

We start from the basics in each chapter. We know from experience that it is riskier to assume too much knowledge than too little. We do not mean to insult anybody's intelligence by doing this.

We progress this way in the book because this is the way we ourselves like to work in any process of discovery and learning, which is what systems change is. If you know your starting points and opening assumptions, you can always backtrack to them if you get lost. In complex circumstances, it is as important to know how you got to where you are as it is to know your location.

What you have learned on the way there, individually and collectively, is your knowledge. No one can deprive you of it. You are, of course, usually free to share it with others who were not on the journey with you. Indeed, doing this often improves that knowledge.

Taking off-the-peg ideas, such as proprietary 'methodologies', does not permit this navigation and this learning. They deposit you in a foreign landscape, with little or no idea of the terrain or how you got there. Hardly ever do they tell you how to get out again with least damage. Such pre-packaged and often expensive 'solutions' have one advantage – they save people the trouble of thinking for themselves. This is about all that you can say for them.

You will, we hope, also notice that this is therefore not a cookbook. Indeed, it is intended as a 'what to' rather than a 'how to' guide. Cookbooks preach one way of making things, to the authors' favourite recipes. They assume that you have all their preferred ingredients to hand, a full range of utensils and the time to follow their methods. All this is very nice in theory but seldom possible in practice.

Our aim instead is to help you become a good cook. This is someone who can make something nutritious and attractive whatever ingredients and tools are available. He can adapt himself and his methods to the circumstances he finds himself in.[1] A good cook still uses cookery books, but as a guide not a prop.

So it is with managing processes. You are unlikely to be in an ideal organization, with ideal people, ideal computer systems, an ideal business strategy and an ideal management style – however you might wish to define 'ideal'. As with being a good cook, managing organizations is a matter of making the best with what you have. Organizational life is too unpredictable and events too interdependent for 'one best way' answers to be usable.

We have assumed that you, the reader, are a mature, intelligent and educated person, able to make up your own mind about possible choices of action. We also believe you do not need coaching in the elements of management, accounting or marketing. Those sections of the book therefore omit the usual theories and studies;

---

1 We are not being sexist by saying 'he', as we do throughout the book. This is solely for concision. It saves having to use such cumbersome constructions as 'he and she' and 'his and hers', which is wearisome for readers and writers alike. ('They' is simply ungrammatical. English may one day have neuter personal pronouns but not yet.) Please, therefore, read any masculine pronoun as applying to both sexes, unless of course it refers to an identified male human being.

you will get those from general business guides. We concentrate on material we think you will be unfamiliar with.

Although much of this material is not technology-related, neither is it there to be skimmed over. The content still needs thinking about. Looking at the world differently is always hard because it demands change in ourselves. This not the One Minute *Process Manager*.

We have written this book to help you to:

- find out what BPM is about
- understand why it is important to your organization
- help you cut though the prevailing marketing hype and technical mystification
- gain moral support for a commonsensical and well-rounded view of the subject.

Even though the book has been sponsored by Staffware, a company that makes process management software, we hope we have been even-handed in technical and product matters. Our simple aim is to help you make a better-informed decision on whether your organization needs to change its processes and, if so, how it should to do so. Whether you use Staffware's software for this or someone else's – or any at all – is your decision.

John O'Connell, Jon Pyke and Roger Whitehead

Acknowledgements

Like most books, this has drawn on the work of many people, to whom we are indebted. In particular, we are grateful to the following for their generous help and advice:

- Jane Lawrence and Claire Grove of Staffware plc for their patient help with case studies and other material
- All the Staffware employees and partners who provided such an abundance of possible titles for the book and especially to Rui Domingos for his winning suggestion for the book's title
- Layna Fischer, General Manager of the Workflow and Reengineering International Association (WARIA), for permission to make use of the entries for the 2003 Global Excellence in Workflow Awards competition
- Eric Willner and Emily Yossarian of Cambridge University Press, for their encouragement and assistance with the preparation of the book.

The case studies and conversations reproduced in the book are shown with the kind permission of:

- Vinícius Amaral, iProcess
- Jeffrey L. Bauermeister, International Truck and Engine Corporation
- Henk van Dijkhuizen, Rabofacet
- Anita Evans, DVLA, and Chris Haden, Anacomp UK
- Chris Mitchell and Neil Chalmers, Morse Computers Ltd
- Roger Studerus, Canton Zug, and Heinz Lienhard, IvyTeam
- Yong Suk Jang, POSCO, and Stuart Kenley, Fiorano Software
- Attila Szász of Hungarian Telecommunications Company (Matáv) and Ferenc Toth Balazs and Zoltan Nagy, Fornax Co., and Tamas Soos, formerly of Fornax
- Petra in 't Veld-Brown, St Regis Paper Company Ltd.
  We also thank:
- Neil Hudspeth of Metastorm Inc. for help with figures 3.4 and 3.5, which we reproduce with permission
- Mike Pearce of IDS Scheer UK Ltd for help with figures 3.7 and 3.8, which we reproduce with permission

- Dan Ternes of TIBCO for help with figure 4.3, which we reproduce with permission.

The reference model shown in figure 5.6 is reproduced with the permission of the Workflow Management Coalition (WfMC.org).

Our grateful thanks go to Richard Holway and Robin Milner for their kindness in contributing the forewords to this book.

Prologue: Olympic results

Every spring there takes place in the USA a sort of Olympic games for users of process management software. Called the *Global Excellence in Workflow Awards*, this international competition has run since 1989. It recognizes those installations that, in the judges' opinion, have been the best that year.

Each entry consists of a case study, which must answer set questions. Several of these questions deal with the effects of the project. We list below some of the answers from the 2003 finalists. We have grouped the first set into three categories – financial improvement, better work management and better competitive position.

Quantified results

Financial improvement

- $2 million a year saved though fraud elimination
- 68 per cent improvement in cost efficiency for the organization
- Paid for itself in three months
- Unit cost of loan 28 per cent lower.

Better work management

- 20,000 hours telephone time saved annually
- 40 per cent increase in on-time installations
- Cycle time down 75 per cent
- Rework down 33 per cent.

Better competitive position

- 90 per cent of expected service levels met
- Customer retention capacity increased by more than 10 per cent

- Increased customer satisfaction (on-line versus telephone inquiries) – 85 per cent. Impressive as these outcomes are, what caught our eye were the *un*quantified results, such as these below. Notice the new category, better information:

Unquantified results

Financial improvement

- Significant saving in 'inventory attrition'
- Staff retention improved, reducing recruitment and training costs.

Better information

- Access to repair statistics allows managers to view trends and to react accordingly
- Customers say the company knows their building schedule as well as they do
- Most important advantage is staff's visibility of processes
- Quick and detailed information available to customers
- Managers feel greater accountability.

Better work management

- Eliminated a backlog of policy applications
- Formalized rules for processing claims
- Immediate implementation of process changes
- Sales staff can more efficiently schedule special offers and other tactics
- Workloads now shared across departments.

Better competitive position

- Allows [organization name] to begin moving to its target operational model, servicing clients in their preferred location, language and time zone
- Faster loan processing than competitors
- Greater customer satisfaction and loyalty
- Improved dealer satisfaction – making it harder for competitors to take away revenue-generating customers
- Now a 'virtual hub', able to take in work from other organizations
- Some customers now take [the company's] information as input to their own processes.

Looking closer at the results

When you compare the two sets of outcomes, there are several notable differences. First is the lack of quantified gains from better information or better delivery of it. Either this is because quantifiable gains did not arise or because people did not have the tools to measure them readily.

What also stands out is the number of qualitative reports of improvements in competitive position. These far outnumber the quantified reports on this topic. Perhaps a lack of suitable measurement tools is the problem here, too.

The other major difference, related to the above, is in the type of benefit these organizations enjoyed. Almost all the quantified results are about doing things cheaper or better. They are aptly reminiscent of the Olympic motto of *Citius, Altius, Fortius* or, more familiarly, 'Swifter, Higher, Stronger'.

This is typically what process management software has always been used for – not changing what an organization already does but performing existing processes more slickly. This is useful and praiseworthy but, frankly, does not warrant reading or writing a book about.

But there is more. If you look again at the unquantified results, you can see some other outcomes and implications, such as:

- More congenial working arrangements ('Staff retention improved')
- Better information ('Access to repair statistics' and 'staff's visibility of processes')
- Closer relations with customers ('company knows their building schedule')
- Improved motivation ('Managers feel greater accountability')
- Order achieved ('Formalized rules for processing claims')
- Greater responsiveness ('Sales staff can more efficiently schedule')
- Competitive lock-in ('Improved dealer satisfaction' and 'input to their own processes')
- Strategic innovation ('Now a "virtual hub"')

It is these sorts of result – less mechanistic and often outwardly focused – that are at the heart of process management today. They show there is room for a fourth Olympic quality – *Ingeniosius* or 'Smarter'.

For business processes, therefore, the Olympic qualities to strive for are *Citius, Altius, Fortius, Ingeniosius* – 'Swifter, Higher, Stronger, Smarter'. It is to help you achieve these ideals that we have written this book.

What we can learn

If you add these two lists of gains together, the result is a resounding endorsement of the abilities of process management systems. It reads like the perfect supplier's brochure.

These results are from just the sixteen finalists for the 2003 competition. There were many similarly successful organizations whose cases did not make the finals. Thousands more did not enter. If all these organizations can achieve these sorts of results, then so can yours.

It is not the suppliers alone who create success, of course. These are users' stories. They reflect their effort, skill, perseverance and courage. They also reflect their vision of what can be achieved by understanding and improving processes.

As the finalists' results reveal, sometimes that vision became clear only with hindsight. One of our main hopes is that reading this book will make such surprises less frequent. We want to help you see your organization's processes more clearly. This will help you more accurately *fore* see the likely results of changing them. We start in chapter 1: A gentle introduction to systems and processes.

1 A gentle introduction to systems and processes

Introduction

This chapter presents some of the major ideas in thinking about systems and processes. It does so by looking at two non-business topics – the Marble Arch, in London, and apple trees. We take the discussion out of the business context for two reasons:

- to help you think afresh about processes and systems and how they should be managed
- to show that real systems are more complex and variable than computer specialists and software salespeople would sometimes have you believe.

Please use the chapter as a refresher if you are familiar with systems thinking. If you are not, we invite you to use it as an introduction to the subject.

Looking for processes

The first vows sworn by two creatures of flesh and blood were made at the foot of a rock that was crumbling to dust; they called as witness to their constancy a heaven which never stays the same for one moment; everything within them and around them was changing. (*Oeuvres romanesques*, Denis Diderot (1713–84))

Much management thinking and writing is about entities – things – that are unmoving, unchanging and separate. The reality is that most of what you see around you, whether you can touch it or not, is part of some process or other. It is on its way to being something else.

Some entities might perhaps look permanent but this is only because the changes they are undergoing are invisible to normal view or are imperceptibly slow. As Diderot suggests, nothing in this world is unchanging.

We want to encourage you to see more of the processes that surround you and to be able to place them in a wider perspective. To get you into the swing of it, we examine two examples from outside the world of business.

Marble Arch

Our first example is Marble Arch in London (figure 1.1). The engraving above shows it as it was over 120 years ago, when the arch was about fifty years old. The arch is still there today, as millions of visitors to London can testify. Looks permanent enough, doesn't it?

Appearances deceive. Not even the arch's position is permanent. It originally formed the main entrance to Buckingham Palace, nearly a mile away. When the palace was extended in 1851, the arch was moved to its current site at the top of Park Lane. It now causes such an obstacle to motorized traffic that there is talk of moving it again.

Figure 1.1 Marble Arch, 1880

What you see when you look at this edifice is merely the present stage of a combination of processes.

The longest of these is a *geological* process. The arch is made of the famous white marble from Carrara in Tuscany. This began as limestone, a sedimentary rock laid down on a seabed 150–200 million years ago. About 100 million years later, the same intercontinental forces that were pushing the Alps skywards subjected the limestone to intense pressure and heat. This changed the rock's internal structure, producing the soft glassy result so prized by architects and sculptors.

The rock's story has not ended. Wind, frost and rain are slowly wearing away the stones of the arch, as they do any rock. The acid in raindrops is speeding this erosion by reacting with the calcium of which marble mainly consists. These granules and chemicals are blown and washed away, being deposited elsewhere. There, they make a small contribution to some river or lake bed, perhaps, that after more aeons will become new rock.

The second process, many times shorter, is an *historical* one – quarrying the stone. People have been extracting marble at Carrara from before Roman times. What you also are seeing therefore, as you look at the Arch, is the output of an industrial tradition over 2,000 years old.

The third process is *human*. It is the career of its architect, John Nash. This is shorter still and lasted about sixty years. Nash lived from 1752 to 1835, beginning his architectural work in the 1770s. His career culminated in a contract from King George IV to develop the then Marylebone Park in west London. This project took nineteen years and resulted in such landmarks as Regent's Park, Regent's Street, St James's Park, Buckingham Palace and, of course, Marble Arch. Nash built this in 1828, modelling it on the Arch of Constantine in Rome, which dates from 313 CE.

Figure 1.2, not to scale, shows today's Marble Arch as being at the intersection of these processes. (MYA means millions of years ago. BCE means Before Common Era and CE means Common Era, in which we live today.)

There are other processes under way besides these three. We have already touched on the development of road traffic management schemes in London. Another is the growth of the public arts around this time. John Nash was one among several architects whose careers flourished in the Georgian period. Others included Robert Adam, James Gibbs, Nicholas Hawksmoor, Sir John Soane and John Vanbrugh. All helped create the neo-classical style characteristic of the period. The development of 'schools' in design or music is a well-known process, as is the way they influence future generations of artists.

It is no coincidence that London was at the time the wealthiest city in the western world. Artists need patrons. There was clearly a process of vigorous economic development under way. Most of the great houses these architects designed were at the time country houses, but not for long. London was also undergoing a process of urbanization.

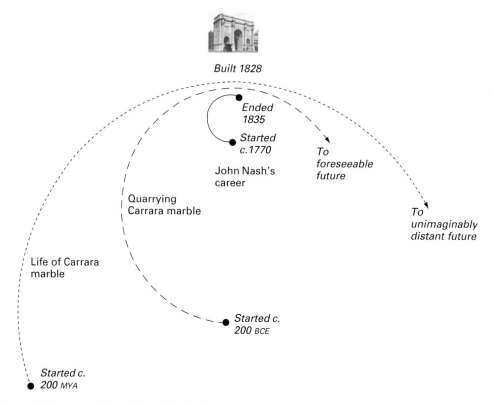

Figure 1.2 Processes at the building of Marble Arch

Selective vision

The more you think about it, the more dimensions and processes you can see intersecting during this period, as you can any period. We have touched on seven – geological, historical, biographical, transportative, artistic, economic and developmental. Deciding which of these is to be representative or definitive is difficult. Can any single process do so? Clearly not, in this case. It is curious, then, that this question so seldom arises when looking at business.

The processes people notice in any situation will depend largely on their responsibilities, background, training and inclination. The three processes we charted above are, for instance, those you might expect a geologist, an industrial archaeologist and an architectural historian to concentrate on.

There is no reason someone from any of those disciplines should not think about the other two processes as well. Surprisingly, though, people's jobs and professions often blind them to other facets of a situation, to other ways of looking. You will see examples of this as we progress through the book.

It is remarkable how often people put on these wilful blinkers when viewing the systems and processes of organizations. It is a trap we wish you to avoid. The next example should also help with this.

Core processes: a story of apples

Fritjof Capra and two Benedictine monks ... suggest that we must see things as processes rather than structures. For example: a tree is not an object, but an expression of processes such as photosynthesis, which connect the sun and the earth. The same thinking applies to our jobs, our organizations, and ourselves. (*A Learning Organization*, Ronald Bleed)

Or, as an evolutionary biologist might put it, a tree is a seed's way of making more seeds. Figure 1.3 shows the outline process.

This looks simple enough. Seeds ('pips') grow into trees, which flower. The flowers turn into fruit, at the centre of each are more seeds. This is repeated indefinitely.

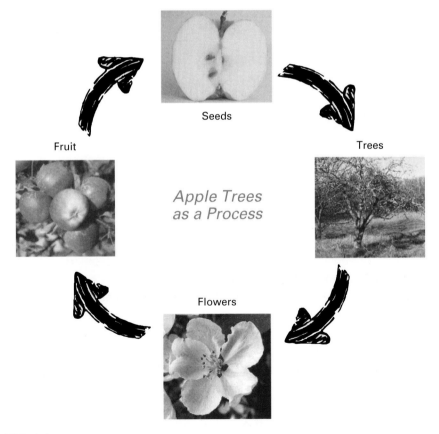

Figure 1.3 Apple trees as a process

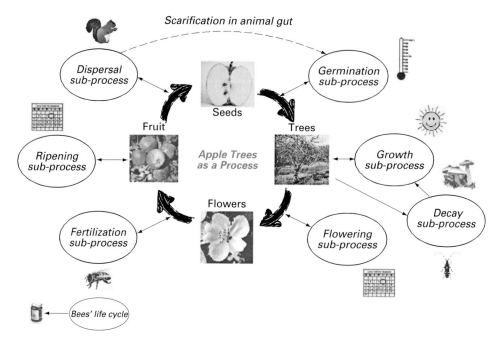

Figure 1.4 Sub-processes within apple trees

Where does the process start? Is it with the seed, the tree or somewhere else on the cycle? It is a chicken-and-egg riddle that has echoes in many business processes.

Deciding start and finish points for computer systems used to be easier than today – your organization's boundary placed a natural limit at either end. Now, as the 'Olympic' cases in the Prologue show, automated processes go into customers' and trading partners' organizations as well.

Business processes have always done this but the means to manage them with the aid of software did not exist before. This is one of the opportunities that BPM offers.

Getting beneath the skin

If we look closer at the apple trees' process, we find that it includes several sub-processes. Most of these depend on some external condition or agent. Figure 1.4 highlights the main ones.

Those sub-processes are shown in table form in table 1.1. Some of them vary in other species of tree.

Don't worry – we are not trying to turn you into an ecologist. This is the greatest depth of biological detail we are going into.

The point of this table is alert you to the number and variety of the processes taking place even in apparently simple systems. It is also to make you aware of the important role that external conditions and actors often play.

Table 1.1 Sub-processes within apple trees

Supplier	Input	Process			Customers
		Sub-process	Determining factors	Output (or result)	
Sun; soil	Sunlight; groundwater	**Germination** (once only)	Temperature, time (day length) and water supply	Sprouting seeds	Tree
Sun; atmosphere; soil (aided by fungi and bacteria)	Sunlight; CO_2; groundwater; chemicals from soil	**Growth** (repeated)	Light, carbon dioxide, water and chemicals	Thickening and lengthening of trunk, production of leaves, etc. O_2 exported in day; CO_2 at night Dead leaves dropped in autumn (contribute to decay process)	Tree
Sun; soil	Sunlight; groundwater	**Flowering** (repeated)	Time (day length) and water supply	Pollinator attractors, giving out scent and visual signals	Tree
Pollinators (mainly insects, esp. honey-bees)	Pollen from other apple trees	**Pollination** (repeated)	Season of year + openness of flowers	Nectar collected by pollinators Own pollen exported Fertilized ovules, needed for fruit production	Tree
Sun; soil	Sunlight; groundwater	**Ripening** (repeated)	Temperature, time and water supply	Fruit Ethylene gas (promotes ripening in other, nearby fruits)	Tree
Birds and mammals (including man)	Semi-digestion of seeds	**Dispersal** (repeated; outside process)	Luck, in the wild, commercial decision in cultivation	Falling fruits made available to outside agents Viable seed deposited on ground	Tree and its offspring
Fungi, worms, beetles, bacteria, etc.	Recycling of tree's constituents	**Decay** (once only; stops all other sub-processes)	Age and health	Dead wood (food and raw material for recyclers)	Recyclers Other plants and trees

In systems terminology, the apple tree's processes are part of an open system. This means the system's boundary can be crossed. Business systems are nearly always open, too.

The layout of the table might be familiar to you if you know about Six Sigma quality programmes. It is a variant of the SIPOC form that quality teams use in the measurement phase of process improvement. (SIPOC stands for the Suppliers to a process, their Inputs, the Process under investigation, its Outputs and the Customers for its outputs.)

Even in this simplified view, there is no single determining factor for many sub-processes. Most of them involve at least two factors: they are multivariate. This is true in business as well, although there is often a temptation to regard them as being dependent on one, especially when assessing performance. (See 'Measuring the results', in chapter 7.)

There is a further link between these natural processes and business activities. Much of the theoretical work on complex systems has been done by biologists trying to relate what they see in nature to other kinds of system. The results of their work lie behind most well-found modern approaches to business process design and management. We shall talk more about this in later chapters.

Death where is thy sting?

The decay sub-process is another important stage in a plant's life cycle, and not only at its death. Recyclers on and in the ground, such as bacteria, fungi and worms, do their work every year on its fallen leaves, its flowers and its fruit. This returns the nutrients in them to circulation, helping the tree recoup some of its outlay. The cost to it of growing leaves and fruit is high.

Decay is a useful metaphor for what also happens in organizations. Old computer systems, obsolete production lines and 'life expired' car fleets do not just fade away. They, too, must be disposed of, usually at a significant cost. Often the law demands that they be recycled. Material recovery is big business. As in nature, recyclers carry out a vital task, often unrecognized. It is easy to forget to plan for the 'death' of a system's components when building a system. This is especially true where those components are people.

Drawing the line

Figure 1.4 shows the difficulty of defining system boundaries. We show a bee and a jar of honey at bottom left of the diagram. Honey-bees visit flowers for their pollen and nectar. Pollen is food for the bees. The nectar gets turned into honey in the beehive, from where humans can extract it for their own use. Are these insects part of the tree's system, therefore?

Box 1.1 Points of view

Ask an apple tree what it thinks of bees and it might say something like: 'Yes, very useful little creatures. They help me fertilize my flowers and swap genetic material.'

Ask a bee what it thinks of apple trees and its reply might be: 'They're a wonderful idea! Filling stations and supermarkets combined – and so many of them.'

Ask a worm what it thinks of apple trees and it would probably say: 'What's a tree?' Explain trees' contribution to soil ecology and it might say: 'Oh, so that's where all that good stuff comes from.'

Ask the worm its opinion of bees. Be prepared for an uncomprehending look even after you've explained what they do. 'Sorry, friend. I'm only concerned with real-world issues.'

It takes just a small shift in viewpoint to see the supply of apple flowers as an input to the honey-bees' life cycle. Should we regard the tree part of the honey-bee's system, instead? Or as well (box 1.1)?

People often do not realize how dependent apples are on those bees. These insects are by far the most common pollinators in orchards. Without enough of them, the trees do not produce much fruit. This is why apple growers place beehives in their orchards, not for the honey but to increase the chance of a good harvest.

There are companies that rent out occupied hives for this. They get a fee, and the honey. This is equivalent, in corporate jargon, to business process outsourcing or BPO.

Sensing and communicating

Apple trees flower roughly at the same time in the wild, usually in May in the northern hemisphere. This makes it easier for bees to find open blooms on different trees, cutting their travel costs. It is also essential for the trees' cross-fertilization.

The trees can synchronize their blossoming because, like almost all living organisms, they can tell the time. They contain the molecular equivalent of a clock that works roughly on a 24-hour ('circadian') cycle. They correct this clock by checking its 'readings' against the daily cycle of light and dark.

While doing this, the apple tree's clock compares the length of nights to the length of days as the season progresses. When the ratio between the two is favourable, the tree releases an internal hormone that triggers the budding and, later, the opening of its flowers. Further on in the year, the shortening days are a signal for leaf drop. Timing, sequence and synchronization are important features of business processes, too.

On being a system

Any persisting pattern of activity that can be described as a system must involve processes that hold it together; otherwise it would tend to degenerate. So, the structure and process of a system and the control of the system are two sides of the same coin. (*Systems, Management and Change: A Graphic Guide*, Ruth Carter, John Martin, Bill Mayblin and Michael Munday)

Making and distributing internal hormones is a tree's way of regulating its growth (size) and development (structure). It is how the tree exercises control over itself.

This sort of responsiveness and internal communication is what distinguishes a system from an assemblage. A tree is a system, whereas a chest of drawers is a collection of inanimate lumps of wood. A more formal definition is that *a system is an entity that maintains its existence through the mutual interaction of its parts*. This is normally attributed to Ludwig von Bertalanffy, one of the biologists mentioned earlier.

Mutual interaction implies both life and communication in whatever is interacting. Dead things tend not to communicate, except perhaps through mediums.

A colony of honey-bees certainly interacts. As is well known, bees can communicate with each other symbolically as well as directly. They pass on details of the direction and distance of a suitable food source through a 'dance' they perform back at the hive.

Our apple trees can also be said to communicate with bees, through their blossom. By appealing to the bees' senses of sight and smell, the trees advertise their flowers' nourishing and tasty content.

We therefore have two separate systems – an apple tree and a colony of honey-bees. Each can by definition communicate within itself. They also communicate with each other. The point where they do so is the flower. It is the interface between those systems.

Process (re)engineering

Another external dependence, shown at the top of figure 1.4, is on the animals and birds that eat an apple tree's fruit. Like many seeds, those of an apple will not germinate and produce a new plant until they have passed though a digestive system. The acid in an animal or bird's gut weakens the seed's hard coating. This effect, called scarification, allows the embryo inside the casing to burst through it in the spring.

One benefit of this for the tree is that by the time its scarified seeds reach the soil, they are mostly far away from the parent. This increases their chances of survival and, thus, of the dispersal and continuation of tree's genes.

This is how it happens in nature but is too hit-and-miss a method for commercial fruit growers. In systems terminology, it is probabilistic. In fact, relying on nature is impossible with many cultivated varieties of apple. They would revert to their earlier, less-differentiated forms if allowed to reproduce freely. (Apple nurseries, as opposed to fruit growers, sometimes encourage this. Most popular varieties of apple arose from chance hybridization.)

Instead of relying on nature's way, growers short-circuit the dispersal and germination sub-processes with something more certain. When they want more trees, growers take or buy cuttings from healthy trees of the required variety. They then graft them on to a different tree, called a rootstock. With tongue only slightly in cheek, you might say this disintermediates the animals and birds.

Something guaranteed to happen every time is called deterministic. Such systems are rarely found outside textbooks and system specifications. In real life, just about every system is probabilistic to some extent. Even grafting does not work every time. It is just more reliable than relying on the birds and the bees.

The farmer's intervention in the natural cycle ensures that the genetic code of the desired apple is followed exactly in its offspring, a procedure called cloning. It is the equivalent of creating and enforcing business rules within organizational processes.

Cloning applies also to the rootstock. Four-fifths of the world's apples grow on a kind developed in the early 1920s at a research station in East Malling, in Kent, UK. This has been cloned repeatedly to produce millions of genetically identical trees and is a recognized international standard.

Routes to market

Assuming our apple grower to be successful, he will need to decide how he is going to sell his crop. What will be his business model (figure 1.5)?

We show four possibilities. For simplicity's sake, we have not shown despatch overseas. We have also assumed that these are dessert or cooking apples, rather than being grown for pulp, juice or cider.

Growers can choose any of these paths to the consumer, or any combination of them. We have shown path A as branching into A1 and A2. A small-volume grower – a 'boutique' farmer – could choose either at short notice. Paths B and C are mainly relevant for medium-to-large farms. Deciding between these would be done infrequently, perhaps only once.

Each flow chart shows the growing process on the left, with the subsequent actions (rectangles) and the delivery and handover points (ellipses) to the right of it. Each rectangle or ellipse is a step or node in the flow. Each handover point is an interface, such as to a wholesaler's system or a consumer's.

The small pictures represent the form of transport used between each node. They give an idea of the volume transported at each stage. In each model, we have assumed that the consumer drives to the place of sale. We have also ignored importing, which is in fact a major economic activity.

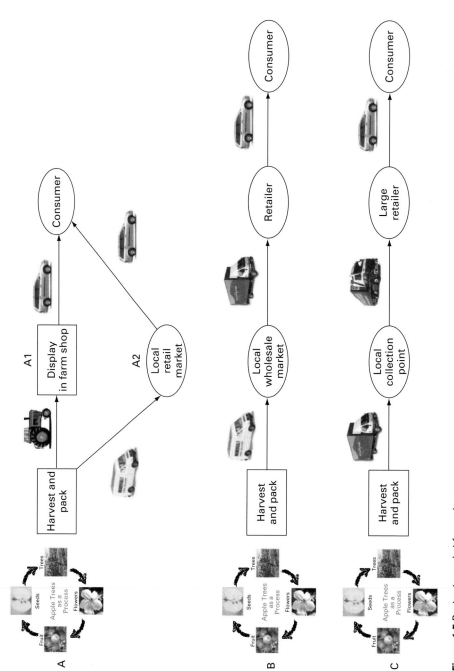

Figure 1.5 Routes to market for apple growers

Table 1.2 shows some of the factors that would possibly influence a grower's choice of how to get his crop to the consumer.

The four main stakeholders in this system are the grower, the consumer, the retailer and society. Wholesalers have a stake, too. The stars are a rough measure of what each path means to each stakeholder. The more stars the better.

We have represented society by the 'food miles' label. As its name suggests, this is a measure of how far a food has to travel to reach consumers. It is, indirectly, a measure of the environmental cost of getting it to their homes. Public concern and national legislation are making food producers pay more attention to such matters, so it warrants a place in the analysis.

Table 1.2 is especially helpful in distinguishing path B from the similar-looking path C. Flow diagrams tell only part of the story of a process.

Table 1.2 also shows that modelling can be done symbolically, using stars in this case, as well as graphically, by flow charts. Mathematics is another common form of modelling, such as in spreadsheets.

Modelling

As with any model, the table above is a simplification of reality. In practice, the grower's decision would take into account many other factors, some of which would change faster than others. The varieties of apple he grows and the size of the crop, for instance, might incline him to prefer one path to the others. So, too, might contractual terms. Other contributory factors would include the quality of his apples, prevailing market conditions, government grants, fuel prices and the long-term weather forecast.

Taking due account of all these is a complex matter, which old-time growers did in their heads or on the proverbial back of an envelope. Modern growers use

Table 1.2 Factors influencing the choice of route to market

	Factor				
Path	Convenience to consumer	Cost to consumer	Profit for self	Profit for retailer	Food miles
A1 (farm shop)	★	★★★	★★★	−	★★★
A2 (local market)	★★	★★	★★★	−	★★★
A3 (wholesaler)	★★★	★★	★★	★	★★
A4 (large retailer)	★★	★★	★	★★★	★

Note: − = Not available.

computer software to model the behaviour of the markets they are entering. This 'decision support' software can guide the grower on what to do but does not direct or manage those actions.

Sometimes these programs are part of what is called a 'full farm package'. This will typically also help the grower plan, budget, schedule and oversee most of the activities on his farm. Inside white-collar organizations, the last of these is referred to as business intelligence (BI) or, increasingly, business activity monitoring (BAM).

Plotting more purposefully

You may have noticed that the life cycle for apple trees in figure 1.3 works only if you deal with trees in the plural. Although a single apple tree can fruit and flower many times over, it germinates and dies only once. If we are interested in the life path of one specimen of an apple tree, therefore, we need to draw the process differently. The flow chart in figure 1.6 shows how this would look for a wild apple if we include the sub-processes depicted in figure 1.4.

The most obvious difference from the earlier charts is that the main flow, that of genetic material, is mostly linear. We start with a viable seed (one capable of life), on the left. The process ends, as you would expect, with the death of the tree. The convention is to call this sort of diagram a life cycle, whether it loops back on itself or not.

There is repetition at the growth stage, so we show that on a loop. Flowering and fruiting appear on it, too. These sub-processes (called sub-routines in computer

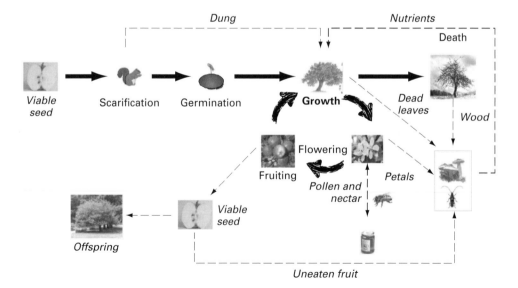

Figure 1.6 Life cycle of a single apple tree

programming) can continue many times over. The original Bramley apple tree, for instance, is over 160 years old and still produces fruit. Most commercial fruit trees last ten–fifteen years before they are replaced.

In systems terminology, each new growth–flowering–fruiting cycle is an iteration. (To iterate is to repeat.) So, too, is each running of a process. Indeed, every apple tree could be described as an iteration of the apple tree process, which takes us back to the quotation about Fritjof Capra.

Each tree is also, to introduce more jargon, an instance of the apple tree process. This simply means that it is an example of the process. An instance (example) can contain several iterations (repeats) of one or more sub-processes.

We show secondary flows in figure 1.6 with dashed lines. These flows are typical of woodland ecology, with recyclers active wherever genetic material is discarded. There are two tangible outcomes of these secondary flows – honey and new trees. (In the wild, the honey would stay in the nests of wild bees, to feed their young, rather than being collected for human use.)

The inward flows in the system as depicted are mainly nutrients recycled by various organisms on or near the tree. These range in size from, say, bears – which are well known for fertilizing woodland – down to bacteria in the soil.

For simplicity again, we have not shown water, carbon and nitrogen. These are, of course, major inputs to the growth processes of any plant. We also do not show sunlight, since that is an energy input. The chart is of the flows of physical material.

One of the smallest inflows in volume – but crucial – is the pollen that bees and other flying insects bring with them from other apple trees. Volume and frequency are not always reliable guides to importance.

Charting conventions

You may notice that some of the labels are in italics, while others are in a roman (upright) font. The labels in italics refer to plant matter, to objects. The labels in roman refer to processes. This is an important distinction in plotting business processes.

You should note that this is a diagram of a physical flow. It does not, for instance, show how energy flows through the system or how information goes through it. Each of these would call for a different diagram (although you could argue that pollen is a source of information as much as of material). When looking at business processes, we need to be equally clear what it is we are plotting the course of.

Also, if we were plotting physical flows within a business, we would start by finding out the size of stocks and the rates of flow of the materials we were interested in. Without these, you cannot get useful information about overall inputs, outputs, throughputs, bottlenecks and so on. Simulation is impossible without this data. We have not done this with the apple tree processes, to keep the diagrams simple.

The apple tree supply chain

Viewed as a business, the apple tree's supply and demand chains can be thought of as containing the external participant shown in table 1.3.

The input from scarifiers is, in effect, part of the factory construction process. (The recyclers of the dead tree are the demolition contractors.) The work scarifiers do on the tree's seeds is an example of outsourcing.

Applying these ideas

That ends the mental aerobics for now. We hope your mind is now sufficiently warmed up to move on to thinking directly about business systems. We will continue to refer to the ideas in this chapter and to natural systems as we go through the book.

Table 1.3 Participants in an apple tree's supply and demand chains

Identity	Role	What it transfers
Sun	Energy supplier	Photons (sunlight)
Atmosphere	Raw material supplier	Gases input to leaves (oxygen and carbon dioxide)
	Waste contractor	Gases output from leaves (oxygen and carbon dioxide); water (mainly as vapour)
Ground	Raw material supplier	Water to the roots
Dunger (e.g. bird; mammal, insect)	Raw material supplier	Food (mainly organic chemicals) to the roots
Recycler (fungi, worms, beetles, etc.)	Raw material supplier	Food (mainly organic chemicals) to the roots
Pollinator (e.g. honey-bee)	Piece-part supplier	Genetic material (pollen received)
	Courier service	Genetic material (pollen exported)
Scarifier	Post-process contractor	Preparation of seed for germination
	Courier service	Genetic material (transported seeds)

Ideas in this chapter

The following list summarizes the main ideas we have introduced in this chapter, and where we did so.

Idea	Page
1. Most of what you see around you, whether you can touch it or not, is part of some process or processes.	1
2. It is usually just the present stage of a combination of processes.	2
3. Deciding which of these is to be representative or definitive is difficult.	4
4. People seldom consider the existence of multiple processes when looking at business processes.	4
5. The processes people do notice depend largely on their responsibilities, background, training and inclination.	4
6. 'A tree is not an object, but an expression of processes' (Ronald Bleed).	5
7. Deciding where a process starts is hard.	
8. BPM lets people manage processes beyond their organization's boundaries.	6
9. Even apparently simple systems can contain many different sub-processes.	6
10. Conditions and actors outside a system often exert an important influence on it.	6
11. Business systems are usually open systems; their boundaries can be crossed.	9
12. Many business processes and sub-processes involve at least two determining factors.	9
13. Building the 'death' of a system's components into the plans for it is sometimes overlooked.	9
14. Defining system boundaries is difficult.	9
15. Renting out occupied beehives to apple growers is a kind of BPO.	10
16. Timing, sequence and synchronization are important features of some systems.	10
17. Exercising control through processes is an essential feature of systems.	11
18. Internal communication and responsiveness to stimuli distinguish a system from an inert assemblage.	11
19. 'A system is an entity that maintains its existence through the mutual interaction of its parts' (Ludwig von Bertalanffy).	11
20. Where systems communicate with each other is called the interface between those systems.	11
21. One way to make a system more reliable is to replace a probabilistic sub-process with something more certain.	12
22. The opposite of a probabilistic system is a deterministic one but these are rarely, if ever, found in real life.	12
23. Cloning genetic material is the equivalent of creating and enforcing business rules within organizational processes.	12
24. Repeated cloning is a way to create a standard.	12
25. Flow diagrams tell only part of the story of a process. You need data as well.	14
26. Processes can be modelled symbolically and by using mathematics, as well as graphically.	14
27. Any model is a simplification of reality.	14
28. Processes can be modelled using computer software. This 'decision support' software can guide the user on what to do but does not direct or manage those actions.	15

2 Business processes and their management

BPM explained: in one minute

BPM is the use of a particular kind of process automation software, typically within the commercial and administrative operations of an organization. This software does five main jobs; it:

- puts existing and new application software under the direct control of business managers
- makes it easier to improve existing business processes and create new ones
- enables the automation of processes across the entire organization, and beyond it
- gives managers 'real-time' information on the performance of processes
- allows organizations to take full advantage of new computing services.

The result is an improved ability to respond to or anticipate changing business demands. Making processes run faster can be beneficial in areas such as customer service. The organization also saves money whenever it changes computerized working methods – usually an expensive and protracted rigmarole. It can extract more value from its existing information technology (IT) investments by putting them to broader and more intensive use. As a bonus, the organization becomes better fitted to exploit future business and computing opportunities, including BPO and Web services (explained in the glossary).

The success of all these depends on how managers introduce and use this new kind of software. BPM is as much about organizational design, human communication, people's viewpoints and mutual consideration as it is about technology. It is not just a matter of optimizing computer programs.

We cover all these matters in the pages that follow.

Oh no, not another book about business process reengineering

You can relax – it's not. Business process reengineering (or redesign) (BPR) achieved some good but has had its day. To its credit, it popularized process-based thinking and the explicit ownership of processes. Against it are its association with job losses, skill depletion and factory-style thinking: a slash-and-burn approach, as some commentators styled it.

Its founders say that people expected unreasonably much from BPR, or did not do it right.[1] Possibly that is true but too many unhelpful or downright harmful actions were taken in its name for it any longer to be credible. One widely quoted article criticized it as 'The fad that forgot people.'[2] There was even a semi-humorous alternative expansion of the BPR initials – 'bastards planning redundancies'.

Justified or not, these opinions were influential. BPR's central idea – 'don't automate, obliterate' – is now damaged goods.[3] Its big bangs too often turned out to be damp – and damaging – squibs.

Organizations no longer want risky, all-or-nothing approaches. They prefer something more sensitive to the needs of the whole business, that embraces continued change and that consequently has a greater chance of success. BPM offers all these.

So it's a book about TQM, then?

No, it is not about that either. Quality management schemes such as TQM and Six Sigma deal mainly with continuous improvement, team working and interpersonal communication. They do not explicitly deal with computer-based process management.

Table 2.1 sets out the major differences among BPM, TQM and BPR. It was inspired by a table in Michael Youngblood's quirkily-titled book on process change, *Eating the Chocolate Elephant*.

Quality marks such as ISO9000, the EFQM Excellence Model and the Malcolm Baldrige National Quality Award in the USA all explicitly include process quality in their scope. BPM software makes it possible to extend the human and technical reach of these ideas and schemes.

---

1 See, for instance, the prologue to the revised edition of Michael Hammer and James Champy, *Reengineering the Corporation: A Manifesto for Business Revolution*.

2 This was also the title of the article by Thomas H. Davenport, in *Fast Company* in 1995. It starts: 'Reengineering didn't start out as a code word for mindless bloodshed. It wasn't supposed to be the last gasp of Industrial Age management.' Strong words, especially as Davenport was one of the founders of the BPR movement.

3 See Michael Hammer's 1990 article in *Harvard Business Review*, 'Reengineering Work, Don't Automate, Obliterate'.

Table 2.1 TQM, BPM and BPR compared

Aspect	TQM	BPM	BPR
Primary objective	Better products and services	Greater responsiveness	Streamlining
Focus	Problem solving	Opportunity seeking	Reinventing
Scope	Micro	All-embracing	Macro
Style	Analytical	Analytical and creative	Creative and destructive
Progress	Incremental	To choice	Dramatic
Duration	Perpetual	To choice	Usually short
Change agents	Whole workforce	Project teams plus rest of workforce	Project teams
Use of computers	Incidental	Fundamental	As an enabler
Control of processes	By managers	By managers	By managers and systems staff

The main constituents of BPM

Figure 2.1 shows what BPM consists of when applied fully.

Moving clockwise from the upper left circle, there are six elements:

- *Process automation.* This is, in effect, workflow automation. Some people feel that workflow automation is not adaptable or versatile enough because of its technical foundations. Such 'religious' disputes are not relevant here, so instead we have labelled this aspect 'workflow +'.

- Making application software work together is a primary task in organizations. This is often referred to as *enterprise applications integration* (EAI) to stress its importance (and, often, to help try to justify its expense). BPM software should be able to integrate any other application software, however labelled.[4] It should also be able to integrate software that runs in other organizations. We therefore call this element 'EAI +'.

- *Business rules* govern the way an organization works. Where these have been written down, in words, diagrams or mathematical formulae, they are often incorporated in the application software that the organization uses. Finding out later what those rules are and making them explicit is hard work. One of the jobs of BPM software is to make that easier, so they can be placed under the direct control of business managers. This is important because business rules change more frequently than the underlying software tasks do.

---

4 Unless we say otherwise, in this book we use 'integration' to denote the sharing of data between application programs. If we mean something else we say so, as in 'business process integration'. So-called 'seamless integration' is usually just marketing-speak. We have never heard anyone own up to seam*full* integration.

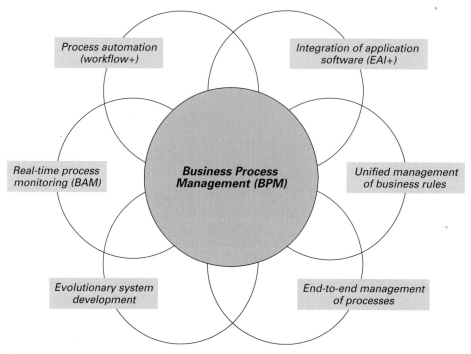

Figure 2.1 The main constituents of BPM

- We mentioned the need to be able to integrate software outside the organization's boundaries. This arises from the fourth requirement for BPM, which is managing processes *across their whole length*. This applies wherever those processes arise and wherever they end.
- An often overlooked aspect is how a BPM system is *created and improved on*. Traditional system design approaches are unsuited to this; adaptive and inclusive techniques are needed.
- Finally, but only because that is how we drew the diagram, comes *real-time overseeing of processes*. The fashionable term for this is BAM. This simply means that process managers see measures of performance and reports of errors or out-of-limit deviations in a very short time. How quickly the manager then responds is his affair but the software usually also allows the equally rapid sending of corrective information.

We discuss all these elements below or in later chapters.

The promise of BPM

BPM potentially offers benefits at every level and in every area of the organization outside manufacturing. It helps people meet these concerns and aims.

Senior business managers   Managing directors, CEOs, general manager and presidents tend not to care about systems integration or the intricacies of business processing. They want to know how the organization is performing. They also want a complete view of the business and to be able to react to changes in market and other conditions.

An adaptable and rapidly reacting ('agile') organization boosts competitiveness. Changes can be made without adversely impacting the business. Post-merger integration is simplified. Inter-organizational working is eased or, indeed, becomes newly possible. Being able to keep computing costs in check is attractive, too.

IT directors, information systems (IS) directors and chief information officers (CIOs)   These topics all matter to them as well, or should do. CIOs must make sure the IT function meets the needs of the Chief Executive Officer (CEO) and the business quickly, effectively and with minimal disruption.

Programming resources go further with BPM. Software maintenance costs fall or are stabilized. Existing 'line of business' (LOB) and other central programs can be dynamically but loosely coupled. Their use can be extended into new areas of the business and their economic working life prolonged. This also eases pressure on budgets.

New systems show fast returns and bring them to the bottom line without having to discard what works. Rapid and modular development becomes possible. Systems and their impact can be simulated before introducing them, reducing the likelihood of rejection. New kinds of external service can be exploited, improving the IT function's business relevance.

Finance directors and chief financial officers (CFOs)   In addition to the above, CFOs want to see closer adherence to plans and budgets. Improved business processes bring this. They strip out waste and unwanted expense, while reducing costly service complaints and product returns. Detailed accounts information and records accrue automatically. The value and working life of existing IT investments are extended.

Marketing managers   Fast and reliable processes make products and services more attractive to potential buyers. They also improve relations with customers, as does supplying them with real-time process information. Processes can form the basis of new services and digital products.

Trustworthy delivery forecasts are possible. Incomplete deliveries, product returns and customer complaints diminish in volume. Common service levels can be agreed and adhered to. Customer self-service becomes possible, safely.

Process managers   Better control reduces unwanted variations in time, cost and throughput rate. Processes run faster (where that is desirable). Faster handling of exceptions and clear problem escalation paths improve service levels, and morale. Responsibilities can be determined more easily.

Being able to control processes directly and change them easily enables more purposeful planning and management. Gaining an overall view of the process improves individual and shared understanding. It also provides a basis for collaboration on fixes and improvements. 'Best practices' can be shared, copied and put to universal use. Changes can be made from a single point.

Employees in general   Processes are no longer dogged with inefficiency, duplication and illogicality. Confusion and frustration decline as a result. People can concentrate on what they are there to do, not on demotivating and time-consuming distractions. Pride in work is again possible. Direct involvement in changing processes or creating new ones boosts self-esteem and commitment.

Nothing is guaranteed

None of these benefits arrives automatically. On its own, software achieves nothing. To gain these benefits, the organization needs to choose a suitable product, apply it where it will make a difference, introduce it well and manage its use wisely.

Before all that, the organization's managers must decide if they really want to make changes. Although these are nowhere near as dramatic or sudden as those demanded by BPR, for example, successful BPM calls for vision, determination and doggedness. Imagination and sensitivity are vital, too. BPM will arouse resistance if seen as a people-expulsion programme.

Also, BPM is not the answer to everything. As with any computing breakthrough, there is a danger of slipping into a frame of mind that says: 'Business Process Management is the answer – what's the problem?' We are enthusiasts for BPM, but not blindly so. We stay in the realm of the possible all through this book.

BPM has to take its place among other systems, technologies and areas of management activity. Also, it applies mainly to the commercial or 'white-collar' activities of an organization. Manufacturing, for example, has its own process management systems. BPM software will or should link to these but not necessarily control them.

The benefits of BPM are not confined to any specific size or type of organization. As the case studies and examples throughout the book show, firms of all sizes, in a wide range of industry sectors, can use it with success. BPM is also not restricted to the type of processes to which it applies. These can vary in scope, tempo, duration and geographical spread.

OK, I'm ready for more detail

At this point, we need to make clear some of our terminology. Like many expressions in commerce and computing, BPM means something more than the sum of its parts. Each of its three constituent words – 'business', 'process' and 'management' – has a separate set of meanings. When put together, the three words have a new meaning, that of supervising and controlling business activities. Thus, business process management includes such well-established tasks as purchasing control, sales supervision and managing a call centre.

By contrast, this book is about Business Process Management or BPM (note the capital letters), which is a particular kind of process automation software and its use. To distinguish it from the general term, therefore, we always show it as we have in the previous sentence. So, when you see 'business process management', you can take it we are referring to supervisory activities of some kind. When you see BPM, you can be sure we are talking only about that a particular kind of software and its use.

It's the business

Another term we should clarify our use of is 'business'. Not every organization is *in* business. There are many other kinds of organization, such as charities, 'non-profits' and those in the public sector. Often, they make or do things of no direct or apparent financial value.

These organizations still employ processes to achieve their ends. Many of these are much the same as those in a profit-earning business. For example, a charity typically recruits people, keeps records about them and pays them. We refer to these activities as business processes, too.

This leads to the definition of a business we use in this book. It is any organization that makes or does things of value to people or other organizations. To create value, these organizations transform raw materials, manufactured goods, services, knowledge and human effort, or simply repackage them. The ways they do this are their business processes. The value arising from them need not, as we have said, be financial or even measurable.

And an organization is?

Two or more people or organizations working together in a mutually accepted relationship. The arrangement need be no more formal than that and can be permanent or temporary. Partnership agreements, contracts of employment and consortium contracts are merely a written record and formal definition of a relationship.

Organizations have no tangible external reality. You can see a company's buildings, its lorries and even its letter-heading but these are just physical pointers to its existence. They are its spoor but are not the organization itself.

What turns a collection of monetary and other resources into an organization is people and how they see themselves. An organization is defined, other than legally, by their and other people's shared perception that it exists. What holds it together are the bonds of duty, habit, economic need, ambition and enjoyment.

Defining 'process'

Conventional practice seems to be to consult a dictionary for this. Bearing in mind that dictionaries describe but do not prescribe, we looked in the two main American and British dictionaries. Merriam–Webster's online version of its *Collegiate Dictionary* offers this:

A series of actions or operations conducing to an end; especially: a continuous operation or treatment especially in manufacture.

The compilers of the *Oxford English Dictionary* give this as the chief current sense:

A continuous and regular action or succession of actions, taking place or carried on in a definite manner, and leading to the accomplishment of some result; a continuous operation or series of operations.

Adding the two together, we get the notion that a process:
- is an action or operation, or a series of them
- is carried out in a particular way
- takes place regularly and is often continuous
- can be subdivided
- is done for a purpose.

That is not a bad start, although we would put the word 'usually' in front of 'takes place regularly'. A business process is not necessarily repeated, whether at regular intervals or not. A company merger, for example, can be a once-in a-lifetime event for the participating organizations.

Another aspect of processes, not always mentioned in business writing, is that they often arise without the help of a single, overall designer. Indeed, some have never formally been defined; they just evolved. As people used to say about BPR: 'Forget the "re-"; simply *engineering* our processes would be a start.'

We therefore prefer this simpler definition of a process:

A sequence of actions and events that, consciously designed or not, aims to achieve a purpose.

A business process, by extension, is:

Any kind of process that takes place within or with an organization or between organizations.

And, finally, 'system'

This is a much-overused and often misused word. If you are to believe their makers' advertisements, it can for instance mean a safety razor, a slimming drink or even a hair shampoo (see box 2.1).

Box 2.1 Our definitions summarized

1. *Business Process Management (BPM)* is a specific kind of process automation software and its use. (This contrasts with *business process management*, which is the supervision and control of business activities in general.)
2. A *business* is any organization that makes or does things of value to people or other organizations
3. An *organization* is two or more people or organizations working together in a mutually accepted relationship, whether permanent or temporary
4. A *process* is a sequence of actions and events that, consciously designed or not, aims to achieve a purpose
5. A *business process* is any kind of process that takes place within or with an organization or between organizations
6. A *system* is any dynamic, as opposed to static, entity whose parts communicate with and depend on each other.

In this book, we use the word to mean any dynamic entity whose parts communicate with and depend on each other. (See chapter 1 if you want more on this definition.) An organization is a system, as is a human being or any living organism. A stock exchange is a system, so is a market. An active network of computers or telephones is also a system.

By contrast, a static collection of parts is just that, a collection. A safety razor is just a handle, a head and a blade fastened together – inanimate and uncommunicative. Slimming drinks and shampoos are just substances dissolved or suspended in water. Although it is normal practice to refer to a single computer as a system (we do so ourselves), until it is switched on it is just an assemblage of components.

Like organizations, systems have their main existence in people's minds. Their constituent parts can often be real enough but the perception that they form a system is entirely man-made. This is also true of definitions, methods, models, classifications, targets, hierarchies, taxonomies, labels and standards. They are not how the world is but how we choose to see it. Often we choose in an arbitrary way.

It is when people forget this artificiality that difficulties often arise. Problems can also stem from one person believing that his view of what makes a system is the same as everybody else's. Miscommunication is the inescapable result.

If that person further believes that his is the only right view, then discontent can soon arises. In large measure, successful BPM depends on people finding out and respecting one another's perceptions. We discuss this further in chapter 5.

The ideas behind BPM

BPM views the organization as a rapidly responding network of resources and actors. This network extends beyond the company's legal boundaries to suppliers, trading partners and customers. These participants can be anywhere in the world at any moment, using both fixed and mobile electronic communication systems to connect to one another. These days, those systems are reliable, efficient and pervasive. They are also cheaper than they used to be.

One of the fundamental aims of BPM is to allow the end-to-end management of processes. The old saying that a chain is only as strong as its weakest ink is true of processes, too. The handicap most organizations work under is that they cannot manage, or even see, all the links.

Figure 2.2 is a simplified example. Marks, Lewis and Skinner ('ML&S') is an imaginary retail organization. It sells to its customers in three ways – from its own shops, from shops-within-shops in other premises, and electronically. Online shopping is by telephone, over the Internet, from interactive televisions and from electronic kiosks. Deliveries to customers are by courier if direct from its suppliers or, if from ML&S distribution centres, by its own vans.

The dark arrows on the diagram stand for the flow of goods and services between the seven entities shown. Light arrows denote the main flows of information.

Imagine that you work in the ML&S head office and are responsible for timely and correct deliveries to customers. Now try to envisage your difficulties if you

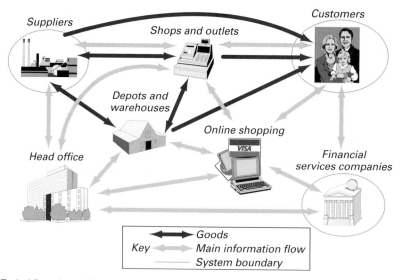

Figure 2.2 Typical flows in retailing

cannot control customer shipments from the depots. Also, you get only daily reports of deliveries made direct from your suppliers. To add to your problems, the financial services companies send you details of customers' online transactions just once a day, for the day before.

The results of this combination of delays and deficiencies are easy to foresee:

- The proportion of incomplete deliveries and returns will increase as more online business is transacted
- If customers change their orders online or by telephone, the only way to act on this is after the (now wrong) order arrives
- Dealing with queries about delays and wrong deliveries diverts staff time from new order taking
- You must either trust that payment for online orders has been made or else hold over them over until the transaction statements arrive.

These problems feed on themselves, with unpleasant consequences. You spend all your day on the telephone, fire-fighting. Increasing numbers of customers become exasperated and go elsewhere. The potential for fraud and underpayments attracts the auditors' attention. Before long, the company's reputation suffers. You are called to account for all this, although it is outside your control.

It might seem we are piling on the agony here but this is a realistic picture. In fact, large retailers trying to fit online sales channels into old ways of selling have experienced every one of those problems and more. Eventually, the survivors have contrived computer systems and data interchanges that allowed timely coordination around all seven entities. These systems are fragile and expensive to maintain. Changing them is not done lightly or quickly. Accommodating new supply chain methods and new channels to market is hard, if not impossible.

This is where BPM makes its unique contribution. It restores elasticity to these now arthritic communications and control mechanisms. All that painstakingly created business logic is collected into one place. There, it can be refined, replicated and modified without the need for expensive and scarce programming staff. Detailed and minute-by-minute information flows around the entire system. This happens even where control by MS&L is not possible, such as of its suppliers' internal processes. Where control *is* permitted, processes can be managed where needed, locally or centrally. You, the person responsible for customer fulfilment, can if you wish see the state of every order and every item. Where there are problems, you know within minutes. If necessary, you can take corrective action within seconds. All this is supported by detailed, regular management reports and a clear audit trail.

Does this all sound to good to be true? Are we trying to sell you BPM? Guilty on both counts, but only to an extent. As you will see from the cases and conversations we show throughout the book, all these improvements are possible. Much depends on how ML&S goes about its BPM programme. We shall pay return visits to it as we consider different facets of the subject.

BPM's technical pedigree

From a technical viewpoint, the main ancestry of BPM software is in document image processing (DIP) and workflow automation products. These have been in widespread use since the 1980s. (For more on the background to BPM, see the appendix, A short history of process management.)

From these, BPM has inherited process modelling, application integration, process monitoring, and rapid application development (RAD) tools. BPM also includes graphical simulation, of the kind industrial engineers have used for years.

Like its name, BPM software is not just a sum of these parts. It brings together all these elements into a single, integrated set. These manage the whole life cycle of a process. It gives unified control of every stage, from defining, modelling and simulating the process through to the issue and running of the operational software for it.

Managers also get business monitoring tools in BPM software. These graphical indicators give them immediate and up-to-date information on the progress of processes. They let managers identify variations from the expected, isolate bottlenecks and breakdowns and identify their repercussions on the business.

BPM software takes the control of computerized business processes away from existing ('legacy') application software and gives it to business managers. The existing software continues to do its job of carrying out those processes but now does so more responsively and under the control of the business.

Making changes to a process and reissuing the working software for it are under the same control. Managers and other responsible people get easy-to-use graphical tools to allow them safely to design and change business processes. This no longer need involve expensive and scarce computer programmers.

All this gives a consistency and a thoroughness that were not previously possible.

Process-centric computing

Adopting BPM also makes possible a fundamental change in the way we think about computer systems and how they work. It promotes a process-centric view, rather than the data-centric view that has prevailed almost from the beginning of commercial computing.

Packaged application software, such as an enterprise resource planning (ERP) product, represents a 'hard-wired' set of process elements. Although these can sometimes be adjusted and tweaked, this is not easy. Often, the result falls short of what the situation really demands. The organization ends up adjusting the process to fit the software.

These products have been likened to logical cement, able to be moulded until they set. After that, you are pretty well stuck with what you specified at the start.

The alternative to packaged application software is to develop your own. Admittedly, you can tailor this directly to your initial needs, but is more expensive and usually takes much longer. (This is why packages are so popular.) Worse, it is no more adaptable once it is created.

Whether using a package or 'roll-your-own' software, the result rarely covers a complete process from end to end. An organization-wide process typically involves several application programs. To ease data transfers between them, modern packaged software usually comes supplied with integrated connectors. Unfortunately, these tend to further embed processes into the software and the computing infrastructure, increasing the rigidity and fragility of the overall system.

BPM software takes a different approach. It separates the task of managing processes from the underlying application programs, their data and their connections. It also insulates the processes from the physical computing infrastructure, such as the types of network and the location of servers.

The independent process layer

This separating out of process management tasks is sometimes described as creating an independent process layer. This 'layer' is entirely metaphorical and does not actually exist within the computers or their software. It is just a useful way of describing what is going on.

The process layer contains a complete view of all the activities needed to carry out a particular business process. It can manage the flow of these activities across different application programs and different departments and groups of people. (Computers can seldom handle all the steps in a process: human intervention is usually needed.)

Using an independent process layer allows greater value to be extracted from existing investments in application programs, content repositories, data integration tools and, above all, people.

Process thinking everywhere, except . . .

every modern management theory – reengineering, process innovation, total quality management, Six Sigma, activity-based costing, value-chain analysis, cycle-time reduction, supply chain management, excellence, customer-driven strategy and management by objectives – has stressed the significance of the business process and its management. ('BPM's Third Wave: Build To Adapt, Not Just To Last', Howard Smith and Peter Fingar)

To this list you could add lean manufacturing, Goldratt's theory of constraints, the Business Excellence Model, the Baldrige awards, ISO9000 and straight-through processing. Business process thinking is everywhere, it seems. Even where it is absent, regulatory and legal pressures are obliging managers to think in these terms. The requirements of the Sarbanes–Oxley and Check 21 legislation in the USA and Basel II in Europe will concentrate minds everywhere on processes.

Everywhere except perhaps for one place. The computer department often seems to be the slowest function to take action on improving cross-functional and inter-organizational processes. This is strange, especially as computer analysts and programmers make constant use of process diagrams in their work. Perhaps it is a case of not seeing the wood for the trees.

This slowness is often not the fault of the people involved. Most computer departments we see are staffed by enlightened, alert and energetic people who sincerely want to improve the business. Something holds them back. Perhaps it is the way their work is managed or the function's internal structure.

A major part of the workload in any computer department is maintaining and tuning its existing resources. People are running hard to keep up with existing demands. This leaves them too little time or resources to take the 'helicopter view' of what is needed across the organization and to act on it. If nobody senior demands this or puts money into the budget for doing so, it gets left undone.

Without a cross-functional emphasis to their work, computer departments will continue doing what they have for decades. This is to produce and optimize vertically integrated systems for specific functions, so-called 'silos'.

It is only relatively recently that data aggregation across functions has been possible, under the drive to data mining and related activities.[5] Previously, computer departments often had a reputation, not always unjustified, for building data islands rather information causeways.

Process, rather than data, integration is still absent in most organizations. One major exception is banks. Many have successfully instituted straight-through processing and 'corporate actions' management. Even there, they have typically applied it only to their major processes. There are no equivalent sector-wide drives elsewhere.

The nature of off-the-shelf products also presents an obstacle, as we discussed above. This is made worse by the fact that the major software packages are often awkwardly different from one another. Their programming languages and conventions, internal structure, interfaces and support arrangements all vary widely. This is why job advertisements in the computing press often specify experience on

---

5 Data mining is the use of sophisticated software tools and techniques to uncover hidden facts contained in databases. It relies on statistical analysis to detect patterns and subtle relationships in the data. It can then often infer rules that allow the prediction of future results. Typical uses are detecting fraud, analysing customer buying behaviour and investigating credit risks.

particular products, such as Oracle, SAP or PeopleSoft. People versed in working across two or three such regimes are hard to find and expensive to employ.

BPM software reduces and sometimes eliminates the need to call on such deep expertise. Also, once a program's process logic is replicated within the independent process layer, there is no need to get tangled up in the application package's entrails again when introducing new or changed processes.

Another contributor to the IT function's isolation on business processes is its history. When commercial computing began, in the 1960s, the machines, their operators and their programmers typically resided in basement offices. This physical and social isolation from the rest of the organization continued through the 1980s. By then, it arose from the need for special air-conditioned accommodation for the computers. That is no often longer demanded but, these days, security precautions instead keep visitors out.

The effect is the same. Most computer functions and their people still, sadly, seem set apart from their fellow employees. Computer specialists' obsession with technology and habit of speaking to users in jargon do not help bring them closer to ordinary people.

Is BPM right for my organization?

By now you might be saying to yourself that BPM sounds interesting but you are not sure if your organization could benefit from it. You might reasonably expect us at this point to set out some pointers for you. These would list the kind of activity BPM would be suitable for, the sort of results you could expect and so on.

It would be a tediously long list. Yes, we know we said earlier that BPM is not a cure-all but, frankly, it is easier to list the situations where it would *not* be useful. Our advice is not to embark on BPM if your organization has:

- Few computerized activities compared with its competition
- Little or no experience of process automation
- Few or no internal computer staff
- One main application programme
- Business processes that hardly ever need changing.

We suggest that in any of these circumstances you start instead with something less far-reaching and ambitious. This reduces the risk of creating a white elephant. These, as you know, are expensive, noisy and embarrassing. A large-scale failure also inoculates people against trying again.

You would be safer setting up a carefully graduated series of small-scale successes. These would aid learning for everyone and create a positive atmosphere. You could then embark on BPM later, if you thought the time and conditions were right. In

---

Box 2.2 Symptoms of 'corporate cholesterol'

Symptom 1 Every customer status enquiry leads to multiple internal telephone calls
Do staff who deal with customers rely on organization charts, company knowledge or simply luck to track down the status of a policy, claim, complaint or enquiry? If so, it is a sure sign the processes and systems do not support the organization's needs. This problem worsens when new products or channels are introduced through takeover or diversification.

Symptom 2 Training classes focus on three-letter acronyms and system codes
Do the organization's training courses deal more with data entry procedures and codes than corporate culture, customer service or up-selling techniques? If they are, it is likely that inefficient, green screen-based processes are hampering growth and driving up costs.[6]

Symptom 3 Staff make heavy use of procedure manuals and compliance memos
Do employees need to refer to manuals and memoranda every day to do their work? This is a sign that business processes are not geared to comply with company policies or industry regulations. Overcompensating by adding extra stages to a process to check and recheck work only slows things down.

Symptom 4 Pens, post-its and notepads are the most prevalent data capture system
When systems are not integrated, employees have to rekey data from one system into another. Usually, they will note down information, such as a customer identity or account number while they move to another display monitor or switch to a separate application program.

Symptom 5 On-the-job experience dictates the productivity of high-volume operations
When more experienced employees can follow a process much more quickly than newer recruits, it is likely that mental work-arounds have been developed to compensate for inadequate processes. (This applies only to routine work, of course. On-the-job knowledge is essential for managers, designers, researchers, system staff and many other employees.)

---

6 'Green screen' refers to older, text-only computer terminals that typically display green characters against a black background. Along with these goes the need to input commands by acronyms and function keys, rather than by mouse. This does not necessarily mean the underlying system will be hard to use but the chances are that it will be.

Symptom 6 Induction periods are longer than the average tenure

High staff churn is a clear sign of process inefficiency. Typically, the highest turnover rates occur where processes are more complex and cumbersome.

Symptom 7 Work is unevenly distributed

If some employees are overburdened with work while others are often idle, the process is not as efficient as it should be. Workers will be waiting for an earlier stage of a task to be completed before they can play their part. Without the dynamic distribution of work, it is harder to process claims, provide estimates or complete new policies at competitive speeds.

Symptom 8 Long queues (lines) at the copy room

Excessive use of the photocopying machine shows a reliance on paper-based processes. These are riddled with costs and are also slow and error-prone. Documents need to be duplicated at every stage, so that more than one person can work on a task at once.

Symptom 9 Folders, filing cabinets and paper trolleys are commonplace

This is another sure sign that an organization is reliant on paper-based processes.

Symptom 10 Small, remote offices process policies faster than flagship branches

As companies grow, a greater number of specialized departments becomes involved in each process. Handoffs, status inquiries and exception handling increase, impeding the ability to process quickly as people do not know one another or are located in different parts of a building, city or country.

chapter 3 we look at some of the kinds of process software that you might consider using until then.

On the other hand, if your organization is not working under any of the restraints just listed, then BPM is worth your while investigating.

A simple guide to whether your organization needs at least some form of process automation comes in a 2004 survey of the insurance industry by the Exigen Group. It lists symptoms of what it calls 'corporate cholesterol' – poor process flows that clog an organization's arteries (box 2.2). Most of these are applicable outside insurance, so we think you will find them useful whatever industry you are in. Exigen's list of symptoms is a useful diagnostic aid to the need for workflow automation. This is an important part of BPM but, as we saw earlies in the chapter, only a part.

Case Study 1   Canton of Zug

The Swiss Canton of Zug is the main local government organization for its area, working from several offices. It wanted to reduce the cost, time and difficulty of recording changes in employee records. These were based on paper forms, which were processed in varied uncontrolled ways. Also, the Canton wanted to coordinate and speed the distribution of the resulting new or updated details.

The chosen solution was to introduce a Web-based workflow product, initially in a pilot project. This went successfully through to production system in three months. As a result, other administrative activities have been automated and the IT function has adopted workflow to help manage its inventory of application programs. Other departments in the Canton are now moving to process automation and there are plans for process links with other cantons.

This is an example of how a carefully managed first installation can set the foundation for broader and deeper computer-based management of business processes. At the outset, the Canton set clear and relatively modest business and technical expectations. It also involved potential users in the design and prototyping of the system. These measures reduced the likelihood and potential consequences of failure, while making success easily exploitable. One intangible but important outcome has been the adoption of a process-oriented viewpoint among current and potential users.

Industry/Sector	Local government	Location(s)	Switzerland
Annual turnover/income	Budget of Canton	Number of employees	1,200
Type of system	Workflow	Supplier and product	ivyTeam ivyGrid
Number of users	1,200	Time to complete	Five months after evaluating suppliers
Business objectives	Create a unified, paperless personnel record system		
Quantitative results	• Roughly 40 per cent reduction in data-gathering workload • Roughly 25 per cent reduction in work of setting up new employee records • New employees are fully enrolled on their first working day		
Qualitative results	• Work throughput is quicker and better controlled • New employees get better first impression of the organization • Organization is a more attractive employer for qualified personnel		

Business background

Canton Zug, in north central Switzerland, has a population of 100,750 people. It consists of eleven communities, spread over 240 km^2, nearly nine-tenths of which are fields, forests, lakes and mountains. The administration of the Canton consists of seven main management areas, a general administration and the justice department. About 1,200 people work in these areas.

Until autumn 2002, all changes in personnel details were carried out using different systems and many paper forms. Data was captured manually, often repeatedly for different applications. New employees often started their new job without having the necessary arrangements made.

System description

Managers decided in March 2002 to introduce a workflow system. They needed to find a product that would allow rapid expansion of a pilot system. It also needed to offer efficient and adaptable implementation of further processes to cover future needs. The two main processes it was to handle were recording someone entering or leaving employment, and general changes in personnel details.

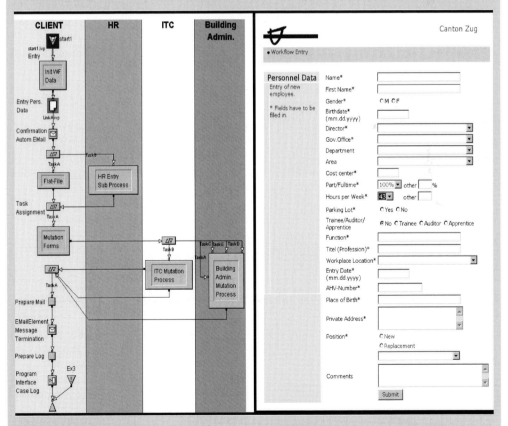

Figure C1.1 Prototyping view in ivyGrid Designer

After a thorough evaluation, the canton selected the Web-based workflow tool ivyGrid from ivyTeam. This allows users to go from a graphical process model directly to the workflow. Another important feature was its ability to integrate other systems into the workflow.

Most of the existing hardware infrastructure could be used directly. New hardware consisted of a Compaq DL 380 server. This runs a Microsoft SQL Server 2000, Windows 2000 Server and IIS Web server, as well as an ivyGrid Server. The workflow software integrates with Personal/400, a human resource management package. Both work into a Novell directory system. ivyGrid Designer runs on a notebook computer. Workflow clients can use their existing browser, such as Internet Explorer.

Figure C1.1 is, in effect, a snapshot of a model in the course of being prototyped. Here, the simulation has just arrived at the process element 'Entry Pers. Data' within the model. This creates the form displayed in the user's browser on the right. When uploaded, the simulation from the designer module becomes the workflow engine of the ivyGrid server.

Implementation experience

Where and when did the project or system originate?	IT service centre, Canton Zug, in March 2002
How long did implementation take?	Roughly three months for the first operations
Who did the implementing (own staff, contractors, etc.?)	Canton Zug staff, with supplier ivyTeam (supplier's staff contributed sixty-seven workdays' effort)
How much bespoke development was there?	None, except for customizing server for specific look and feel
Were there any special infrastructure needs?	No
What were the most significant implementation issues and how were they dealt with?	Integration into existing applications; tool offered straightforward solution
Who is responsible for the system overall?	IT project manager of the Canton
How was and is training handled?	No special user training needed (user interface is a standard browser) One day's training for workflow administrator
What was and is done to encourage use?	Users were involved in process design and prototyping, hence highly motivated to introduce the system
What lessons were learned?	Validation of the ivyGrid design philosophy of going directly from process design to the workflow implementation involving the people concerned. Forms design is also important for user acceptance.

The Canton ran a workshop in May 2001 on the procedures for changing personnel data. Weak spots were identified in relevant processes, especially those involving the human resources (HR) and building departments and the IT service centre. These departments are in different locations.

Six problems emerged during the workshop:
- multiple data capturing on different forms
- uncoordinated circulation of forms, through the lack of a well-defined process
- a lack of information on and for new employees
- generally low quality of information
- unreported personnel changes
- telephone changes often not reaching the building department.

Managers had to register personnel data repeatedly to different offices. As a result, information from the human resource department to other groups often arrived too late, or not at all.

As a first step, the Canton introduced a global directory service. This was to synchronize the personnel data directories of the systems in various offices, such as HR, telephone directory, access control and facilities management.

In a second workshop, the project team established 'should be' change processes and agreed on the importance of controlling and managing these processes. The HR department would record main data, including name, sex, birth date, organization assignment and cost centres. It would simultaneously make these available to all departments involved.

The new processes were distributed for comment and corrections. Managers then decided that in a next phase the processes should be supported by a workflow solution properly integrated into the existing IT infrastructure.

In March 2002, managers sanctioned a pilot workflow project to let the IT service centre gain experience in managing processes.

Because the system goes easily from a process model directly to the workflow, people immediately recognized weaknesses in the original specifications. One outcome of the pilot was therefore an increase from the original two processes to seven:
- enter employment
- enter external contractors (a sub-process)
- leave employment
- personnel data changes (process started by the employee's supervisor)
- personnel data changes by user (process started by any employee)
- change originated by the IT department
- transfer of workplace.

The biggest hurdles to overcome had more to do with culture – possibly national culture – than anything else:
- sometimes people were hard to motivate for the organizational aspects of the project

- the existing matrix organization meant that people working on the project had to perform their main line activities at the same time; this made it difficult to finish the project within a tight time frame: clear task assignments and deadlines were crucial to success.
- the given budget was too small to meet all the new needs and requirements that arose after the first implementation; the positive side to this was that some people got excited about all the new possibilities.

Benefits and user reaction

What has been the reaction of managers and staff?	High acceptability
What has been the reaction of customers or trading partners?	Very encouraging; other Cantons are looking at the system
What has been the overall cost of the system?	Around US$160,000, including licence costs and implementation by ivyTeam
What have been the main process benefits?	Processes are fully supported and thus consistently followed. There has been a significant increase in quality and efficiency
What have been the main effects on operating style and methods?	People now work in a process-oriented fashion

The first seven workflow processes have been running since October 2002. User acceptance is high and significant results have already been obtained. The system is continuously being improved and new processes added. All the communities within the Canton have now been tied into the workflow.

Six key advantages of using the system are:

- wrong process handling is minimized
- omitted process work escalates automatically (that is, goes to a higher level of management)
- data quality has improved significantly
- work has become easier for people involved in the process
- manual tasks are being taken over by the workflow system
- throughput time of the process is significantly reduced, saving taxpayers' money.

Much has changed for direct users and for everyone involved in the processes of changing personnel data. The people ordering or initiating changes now face only one system to register any type of change. All paper forms have disappeared and any information has to be entered only once. Newly hired people find everything ready for them to start work efficiently and even have their car parking space ready from their first day.

People involved in the workflow now receive their change tasks from the workflow system. It also helps them carry out their tasks. For example, a button click gives them a prepared letter in which currently available data are inserted by the workflow. After editing it can be mailed by another button.

Tasks are now well structured and ordered according to the area they belong to. For example, a help desk employee can now enter a new user ID in a simple workflow form. The information spreads through the workflow into the directory and into user administration.

Four key performance improvements are:
- a roughly 40 per cent decrease in the effort needed to get relevant information
- a roughly 25 per cent reduction the effort needed to define a new user
- controlled and shorter throughput times
- a new employee now has the complete infrastructure at his fingertips on his first working day.

Introducing the workflow supporting the core processes has improved the efficiency of Canton Zug's administration. A new employee's first impressions can be decisive. The attractiveness of the administration as an employer for qualified personnel has perceptibly increased.

In summer 2003, the Canton introduced a workflow-based dossier management system for naturalizing citizens. This has proved so efficient and popular that other Cantons have expressed interest in using it.

The two main lessons learned were:
- although the process is at the centre of these applications, one should not underestimate the need for optimal forms design; data entry and presentation often determine how users feel about the system's efficiency and ease of use
- people have to be prepared for possible organizational changes because of introducing workflow automation.

The future

The processes already set up will be continuously improved. Core processes from other areas of the administration will be implemented as workflows in the future. Since February 2003, for instance, the police department of the Canton has been using workflow for some special processes of its own.

ivyGrid can call Web services within a process and lets new process-based Web services be set up easily. There are plans to use these abilities to create 'interworkflows' with other Cantons.

3 The job BPM software is there to do

Introduction

The idea of managing processes has long been familiar in manufacturing and extractive industries. Industrial engineering and work study, for example, date back to the early twentieth century. More recently, ideas such as concurrent engineering and 'lean' manufacturing have emphasized cross-functional and inter-enterprise working. These ideas have begun to permeate the financial and other white-collar sectors.

Quality management schemes such as TQM and Six Sigma have added consideration of continuous improvement, team working and interpersonal communication. Quality marks such as ISO9000, the Business Excellence Model and Baldrige all explicitly include process excellence in their scope. BPM software makes it possible to extend the human and technical reach of these ideas and schemes.

Processes have always been all around us. The difference these days is they are increasingly becoming visible and explicitly managed. Despite this, there are plenty of processes that formal schemes do not or cannot cover. You need to accustom yourself to looking out for all kinds.

This chapter begins by tracing the rise of process thinking and process management tools. We then examine what these tools must do if they to translate organizational reality into worthwhile computer programs. Finally, we review the main types of process management software available to you, pointing out their main areas of application.

Looking at processes

Before a computer can carry out or enact a process, it must be presented with details of that process in a way it can understand. This is not easy. As William Kent, then a database expert at Hewlett-Packard Laboratories, bracingly put it in his book *Data and Reality:*

The art of computer programming is somewhat like the art of getting an imbecile [*sic*] to play bridge or to fill out his tax returns by himself. It can be done, provided you know how to exploit the imbecile's [*sic*] limited talents, and are willing to have enormous patience with his inability to make the most trivial common sense decisions on his own.

The resulting set of instructions or computer code is held within an application program, commonly called an 'application' or 'app'. (In reality, the application is the task the software is there to do but many computer people do not bother with this distinction.)

To write that application program, the programmers need a representation of the process. This can be visual, verbal or numerical. Often it is combination of these. Figure 3.1 shows a fictitious example. It is one of several charts that a system analyst would need to give a complete description of the process (complete as far as a computer is concerned, at any rate).

Unless you are used to this particular modelling method, called Unified Modelling Language or UML, it is not easy to see what is going on here. Mainly this is because UML is for computer software designers, not for users. It may be unified but it is not universal. Figure 3.2 instead shows something less intricate.

This is the kind of flow chart that a work study engineer might have produced in the days of stencils and clipboards. It shows actions (circles), a delay (elongated D), transport (arrow) and an inspection (square). The engineer would number and, usually, time each of these. This data would form the basis of a method improvement program. It would also be used to set a standard time for carrying out the task recorded.

Charts like this – although not of this particular process! – are still in use by industrial and quality engineers. This example is at a much higher level than real charts, which go into great detail. These days, work analysts use charting software for recording tasks. As well as giving a better-looking result, it allows them to make revisions easily.

Although useful for studying the physical movement of workers, materials and tools, this kind of chart is not well suited to examining office work. It does not show information flows, for instance, or decision points. As drawn, this chart also does not permit variation from the process.

The next charting method offers a different style of representation. It is widely used in quality management analysis and goes under various names – functional (or cross-functional) chart, deployment chart and swimlane (*sic*) chart are the most common (the last name is because it supposedly reminds one of the lanes in a swimming pool).

This example, even more fanciful than the one above, shows the imagined waking procedure of a rich person (figure 3.3).

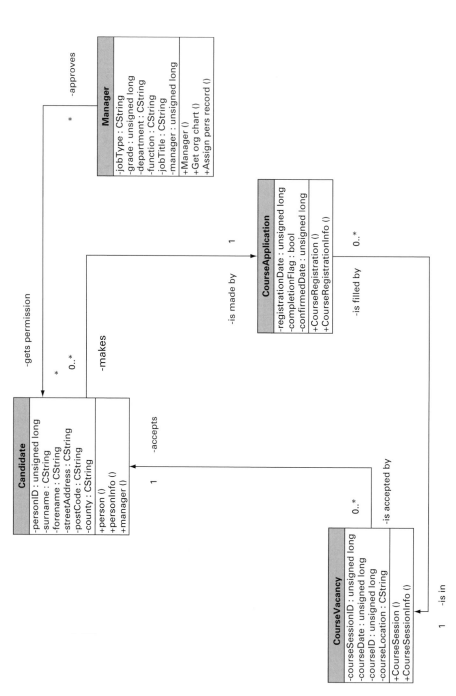

Figure 3.1 Course booking

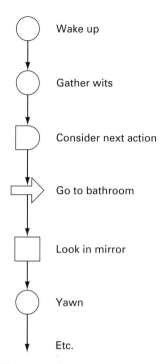

Figure 3.2 A process for waking up

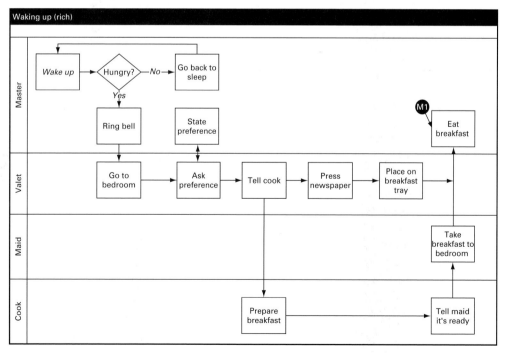

Figure 3.3 How the wealthy awake

Here we have gained useful detail:

- The swimlanes allow us to show the activities of multiple actors. These can be organizational positions, as here, or occupational roles or machines.
- Decisions, shown by the diamond. These are a special kind of action. Processes link points of decision, and link decision-makers. Note that the one shown is a two-way decision (binary). Not all decisions are binary. There could be a third path, out of the top apex, labelled 'No, but expects to be soon' or similar.
- Because of the decision, there are alternative paths. Note that these are of the flow of individual people's actions. One tracing the flow of information, say, would differ slightly.
- There is a feedback loop, allowing the revisiting of an element or step in the process.
- Two flows meet, as the valet places the freshly ironed newspaper on the tray the maid is carrying. We do not know where he intercepts her; the chart does not carry positional information. We can however safely infer from this that there is some method for synchronizing these two sub-processes. The valet can probably either hear or see the maid as she nears the bedroom.
- There is measurement, M1, at the point when the master of the house eats breakfast. This will likely be a qualitative measure ('Delicious, Jeeves. Please convey my respects to the kitchen.'), unless the master of the house is the type of person to award a score out of ten.

Even now, there are details we cannot gain from the chart. It shows sequences, for instance, but no times. We do not know exactly when the steps need to take place relative to one another. Nor do we know if there are target elapsed times, tolerances, deadlines and so forth. Also, there is no cost detail attaching to any step. This is all important information in a more industrialized process, especially the last.

What we need to know about a process

To get a fuller picture, we would need to gather and make available to a programmer the sort of information listed in table 3.1.

We also need to find out whether any of the process detail needs to be concealed from the view of other processes. This is common when a process links to those outside the organization or organizational unit. We would then need to ensure that only necessary details and patterns are visible to external view. Using an intermediate process is one way. This would expose only an interface to the real processes in the company and not the processes themselves. In systems terminology, our process then becomes a 'black box' to those other processes.

An example of this is the information on delivery progress that companies such as UPS and FedEx allow you to call up over the Web. These services tell you what you need to know without giving you any further insight into the courier's internal systems. (Incidentally, would your organization as brave and confident as these

Table 3.1 Data needed for process modelling

Actors	Interfaces to other processes
Boundaries	Mode of action (online, batch or manual)
Branching	Name of process owner
Business rules	Object flows (e.g. documents)
Communication links	Performance measures
Concurrency ('fork and synchronize')	Purpose and aim of process (not necessarily shown on the model or diagram)
Constraints	Resources consumed
Control flows	Roles
Costs	Sequences
Data flows	Start and finish conditions or specifications
Decisions (as described in rules)	Stocks
Delays (and their criticality)	Time (for each activity and elapsed)
Entities (resources, actors, work items)	Timing constraints, such as time-outs and deadlines
Exceptions or faults (asynchronous)	Value adding or not (for the customer)
Feedback loops	Volumes (qualitative or quantitative)
Frequency	
Inputs and outputs	

logistics companies. Would it be willing to show its customers, and possibly its competitors, how well it is meeting its promises? Even if it were, would it be able to do so item by item, in real time?)

Lumping and splitting

Designers find it helpful if the software allows them to aggregate smaller tasks into super-tasks, for speed, ease and reuse. Conversely, they also like to be able to break larger tasks down into sub-processes where possible, for similar reasons.

Figure 3.4 portrays a typical process for taking on a new employee. It was drawn using Metastorm's e-Work product, which is intended for use by managers as well as systems staff.

Even though the result is less daunting than that the UML chart in figure 3.1, it conceals a great deal of complex processing and decision-making. The software hides this from view by the use of ready-packaged sub-processes. You can see them listed on the left of the zoomed-out version of the same screen shot in figure 3.5.

Done thoroughly, process plotting (or mapping) and design is no five-minute task. Important as it is, 'manager friendliness' is not the only facet to take into account when choosing process design software. The design tool has also to

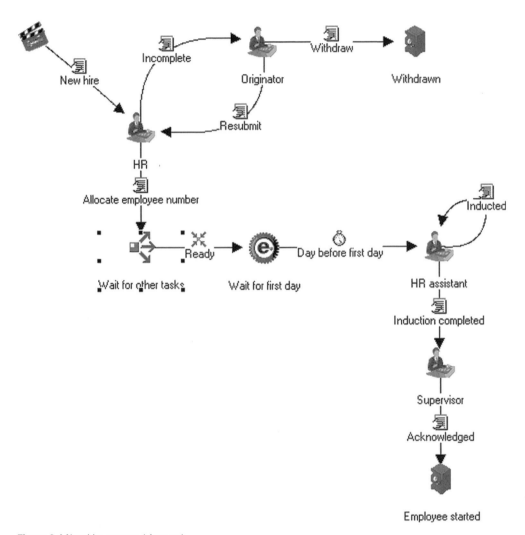

Figure 3.4 New hire process (close-up)

interface with the software that will run the process, producing instructions that it can follow precisely and efficiently. The Metastorm software clearly does this, as the Morse case study in chapter 6 testifies. It is an all-in-one product, with design, enactment, management and other features all from the same maker.

UML-based design tools can potentially produce instructions that a range of process enactment software can use. This is the universality spoken of in its title. Design software can be from a different maker than the enactment software. (This is true also for design software not based on UML.) The decision for any software buyer is whether the organization needs a component-based approach, an all-in-one product or some hybrid of the two.

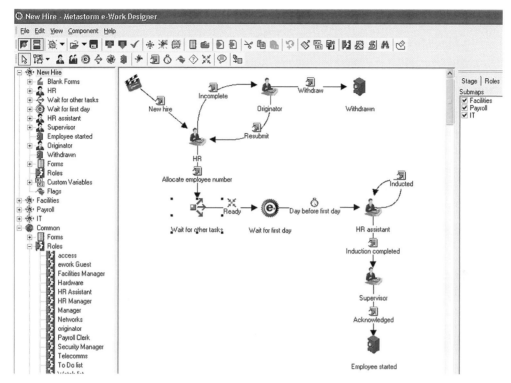

Figure 3.5 New hire process (with design tools)

A digression on standards

Using a combination of separate, specialized products is called – sometimes hopefully – the 'best-of-breed' approach. Its proponents slightingly compare this with using what they call monolithic products.

Continual advances in standards make it increasingly possible to split out the components of what were previously all-in-one products. This is grandly called 'exposing the architecture' and applies in all areas of computing, not just process management. The argument is that this allows specialist suppliers to offer better products for particular parts or layers of the architecture, to the user's benefit (and those suppliers', naturally).

Against this practice is the suppliers' habit of tinkering with standards, diminishing their value. Sometimes this is for good technical reasons; too often it is a kind of competitive reflex. Also, where once a user organization would simply specify and buy a 'monolith', when using components it has to assemble these sets of tools itself. These do not always integrate as well or as easily as their makers sometimes claim. The result, as you would expect when various breeds commingle, is often a mongrel. A popular alternative is to buy these assemblies of specialized tools ready-made, from software suppliers or systems integrators. The headaches and risk are then theirs.

Processes versus procedures

These two words are often used interchangeably – we are no doubt guilty of it in this book – but it is important to distinguish between them. In chapter 2, we defined a process as a sequence of actions and events that, consciously designed or not, aims to achieve a purpose. By contrast, a procedure is a sequence of actions and events that conforms to a set of explicit instructions. A process is goal-directed; a procedure is rule-following. Procedures are consciously designed; processes can arise spontaneously.

Procedures are vital to the running of an organization. Most computer programs embody procedures, for example. Workflow programs also enforce procedures. The major quality management schemes are all based on procedures. Procedures are also the basis of corporate governance programmes. The list is lengthy.

Setting and enforcing procedures and rules can be helpful at the operational levels of an organization, but not always. Some rules or procedures help towards achieving a purpose, implicitly or overtly. As the old medical line has it: 'The operation was successful but the patient died.' Something more aware is needed, more aware both of its purpose and of other systems around it.

It has long been possible to obtain software that learns as it goes. Using heuristics (from the Greek for 'to discover'), these programs work to rules of thumb or guidelines, as opposed to precise and unchanging procedures. They are mostly used to solve problems, such as assessing whether an email message is spurious or a credit rating is representative. Although less deterministic than ordinary software, such programs are not goal-directed, since they must have problems assigned to them.

An objective seeking program can not only recognize a problem or variance and find a corrective to it, it can apply that correction. (This is negative feedback.) It can also optimize resources and actions to better achieve the stated objective. (Positive feedback.[1]) In other words, goal-directed systems take the initiative. Rule-following systems react.

Different viewpoints

As the UML chart in figure 3.1 demonstrates, a single view does not meet all purposes. This is true for any of the widely used charting styles.[2] As a minimum, one would typically want to see separate charts for organizational units, organizational or role structure, data objects, decisions and information flows.

---

1 Positive feedback acts to amplify signals coming in to a sub-system or device; negative feedback acts to diminish or attenuate them.

2 These sometimes bear names as arcane as their internal conventions. They include Bachman, Booch, Chen, Coad/Yourdon, IDEF, Martin, Petri net, Rumbaugh, Rummler–Brache, Shlaer–Mellor and SSADM.

No single charting style can display everything we need to know in one view. Even if it were theoretically able to, the result would be impossibly cluttered for ordinary users. Only highly trained specialists would be able to make sense of it.

Hard-to-interpret charts also defeat one of the other objectives of charting. They put an obstacle in way of using a chart to communicate with people. This is as important as being able to code a computer program accurately. Being able to see different views of a process helps you produce better models. There is always a danger of narrowing down too quickly, of trying to force agreement to an inadequate representation of a process.

Despite all the refinement and ease possible with charting software, the most-used modelling process tools in the world are whiteboards, flip charts and 'the back of an envelope'. They work best when used collaboratively. These low-tech tools do the vital job of helping to get people's ideas and knowledge out into the open and recording them. There are no software licence fees involved and no training needed. Computer charting comes into its own after this stage.

Data-driven charting

Most modern charting software is based on a foundation of data, of the kind listed in table 3.1. Metaphorically speaking, the data sits behind each chart, with links to other data sources and to application programs. You can then choose the level of detail a chart shows, suppressing or revealing detail to choice. Inert pictures cannot do this and are of limited value for modelling; they need to be 'live'.

Modelling programs vary in the amount and range of data they can contain. The charting modules in serious workflow and BPM software should cope with every-thing and more listed in table 3.1.

In figure 3.6, drawn using Staffware's Process Designer tool, you can see the outline flow of a sub-process in the main pane, at upper right. To the left of it are icons representing various other sub-processes. These work with the one shown in the main pane. The person designing the process can insert or remove these as needed.

Below, still on the left, is the catalogue of detailed steps relevant to the process. These are at a finer level of detail than sub-processes.

Steps, or operations, are the basic building blocks of every process. A sub-process packages a recurring combination of steps, to save the designer time. Where there is no relevant sub-process, the designer will combine individual steps to model that part of the process, or create new steps.

Underneath the main diagram are two data tables. These help with designing the screen that users will eventually work to. The tables specify the type and amount of data that goes in various fields on the user's screen. This is the same as designing a database entry form. This is no coincidence – all modern process management software works from a database, sometimes called a repository. It is either integral

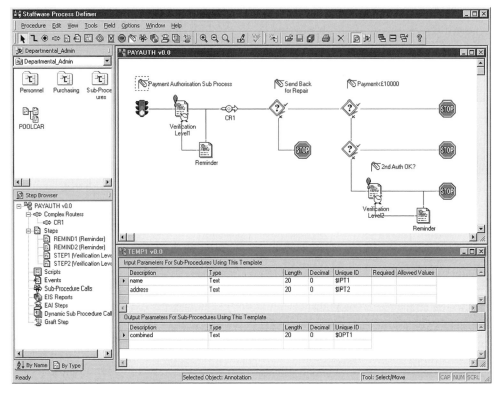

Figure 3.6 A data-driven process design tool

with the process software or is a standard database management product, such as Oracle or Microsoft SQL Server.

Simulation

Software for simulating processes offers similar views to that in figure 3.5 but can also present dynamic information, sometimes with moving pictures. This is done using what are called discrete event simulation tools. These show what is happening during the process, with lights, icons or bar graphs indicating changes in the situation. In effect, the software pretends to run the process design as though it were real. This lets you try out different possibilities, playing 'what if' games. Designing and play are closely related.

Here are two examples, supplied by IDS Scheer and made with its ARIS product. Figure 3.7 shows a moderately detailed view of an imaginary process for issuing an invoice.

Figure 3.8 shows the results of running a simulation of that process. The upper window shows the process in compressed form, with sub-processes collapsed into

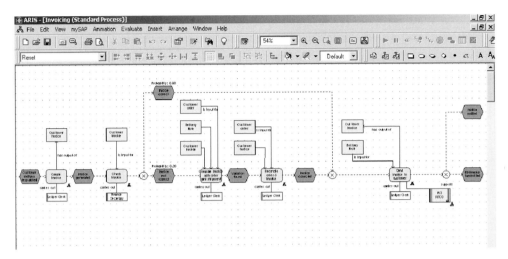

Figure 3.7 Invoicing process to be simulated

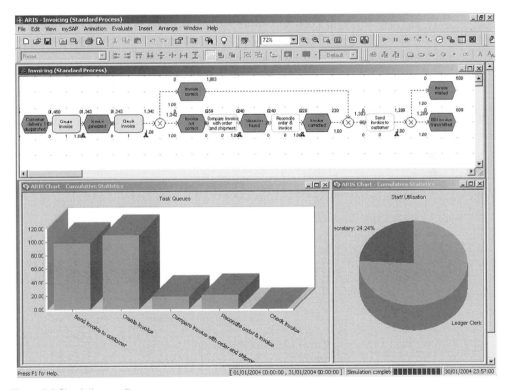

Figure 3.8 Simulation results

the main flow. On the lower left is a window giving in graphical form (but not in sequence) the effects on the task queues for five main actions:

- sending the invoice to the customer
- creating the invoice
- comparing the invoice with the order and shipment
- reconciling order and invoice
- checking the invoice.

In the lower right window, ARIS has produced a graph of how the work content of the task divides between two job positions – secretary (roughly a quarter of the time spent) and ledger clerk (the rest).[3]

You can set a simulator to run a process as many times as you think necessary. Each time, the software shows what the outcomes would be if some of the variables were to change. As well as the results shown above, measures might include:

- the time taken at each step
- number of people working on the process
- variations in the difficulty of each type of work item
- the volume of work coming through.

At the end of each cycle of iterations, the simulation software produces a numerical report. This typically feeds into a spreadsheet program for further manipulation, analysis and experimentation. Once the design is settled, it then passes to the process management system for turning into the production version.

All this, as suppliers of simulation software will tell you, is cheaper and less disruptive than launching untried processes in real situations. They are half right. Simulating a process is undeniably better than doing everything in one go – the 'big bang' approach – but is not the only way forward. Small bangs are often possible.

The use of RAD and similar techniques is becoming increasingly common. Their underlying principle is that designers release incremental draft versions of a new process for users for experiment with. The users try the designs, knowing they do not represent the final word, and go back with suggestions for improvement. This cycle continues until all or a practicable majority of suggestions are incorporated in the design. Only then does the design move into the final stages of becoming an operational system. By this stage, the intended users are familiar with the new process. We shall deal with this topic in more detail in chapter 8.

Simulation can play a part in this learning. By running a series of simulations, designers and users both gain useful insight into the properties of a proposed new system. They can see how it is likely to behave under varying conditions. Often, this behaviour could not be predicted beforehand. Occasionally, the system's responses

---

3 This process is merely an example designed to show how simulation works. If it were real, you would probably wonder why so much time is being spent on checking rather than doing. Invoice-less working, for instance, can be found in many industries. Simulation should go hand in hand with process simplification and improvement.

defy common sense, displaying counter-intuitive behaviour. (Reversing a trailer or caravan is fraught with this, for instance.)

These behaviours are the emergent properties of a system. They are the reason that some days you get result C when you figuratively press button A and pull lever B, while on other days you get results D, E or Z. This characteristic of complex systems has led many computer projects to fail to meet their intended targets and sometimes produce unwanted outcomes. A well-constructed simulation can help forestall these failures. Simulations do not just give an answer, they help you see how the answer came about.

Business rules

We mention business rules in the list of required data shown in table 3.1. This is a popular term with suppliers, consultants and journalists. It simply means 'what people do' or 'what a procedure does'. Here is an example of how the business rules for a (British) milkman might read:

1. Drive to your usual starting spot in this street and stop
2. Get out enough full bottles for the next house on your round, with any extras, such as cheese or bread
3. Take them to the house
4. If there is a note saying no milk wanted today, go to the next house if close enough; cursing is optional
5. Repeat step 3 until household wanting delivery is found
6. Leave goods on the doorstep
7. If there are empty bottles on the doorstep, pick them up and return to the van; put the empties on the van
8. If not, return to van
9. Resume round.

To simplify the described procedure, we have made two assumptions. The first is that either there are no empty bottles at the houses not wanting a delivery that day or the milkman decides to leave them there if there are. Both are unlikely in practice. The other assumption is that all households want the same goods and in the same quantities. Again, both are unlikely.

The reality is that the milkman will be making decisions as he goes from house to house. He must constantly balance the need to minimize journeys to and from the float with the need to keep his customers content, whatever the provocation. This takes skill and practice.

Despite this, a milkman is not someone normally classed as a 'knowledge worker'.[4] His is a low-paid job demanding few educational qualifications, yet it calls for

---

4 The expression was popularized by the management writer, Peter Drucker. He first used it in 1959, in a book called *Landmarks of Tomorrow*. Drucker wanted to distinguish manual workers from those people

judgement, memory (long and short-term), a sense of geography, tenacity, selling ability, reliability, a driving licence and the willpower to get up early in the morning. No computer can match this combination of talents and qualities.

The other matter to note is how cumbersome the verbal description of these rules is. The algorithm becomes harder to follow where there are exceptions and repetitions. A picture like that in figure 3.9 clarifies the options. This is a decision tree, often used in such circumstances.

Like the charts in figures 3.1 and 3.2, the decision tree in figure 3.9 makes no reference to the volume of goods, distances travelled, time expended or other such important information. It is an abstraction of a process, as are they.

When making a model of a process, therefore, we need to use different ways of representing it, depending on what we are trying to achieve. Decision trees, with rough drawings and sketches, are particularly useful in the early stages of process mapping. They make it easier for users to describe to an analyst how they see the processes they engage in. Users can do this in groups, collaboratively creating drafts to be discussed and refined. These drafts can later be turned into the kind of formal statement of rules that process management software can use.

Patterns

Rules can be grouped into patterns. These are sets of typical or recurring processes, made into templates for use elsewhere. They are, in effect, large-scale versions of the sub-processes that graphical design tools make available (see figures 3.5 and 3.6). Instead of offering pre-packaged routines for, say, approving an invoice, these patterns offer templates for complete business processes. These might include such near-universal activities as taking on a new employee or making sales. Equally there are patterns for specialized processes such as settling insurance claims or taking on a new subscriber to a telephone service.

Large general-purpose software suppliers such as IBM, Oracle and Sun all offer patterns, of varying levels of abstraction, as do specialists such as Staffware and Siebel Systems and some standards bodies. Here is how IBM describes its 'Patterns for e-business':

Business patterns are high-level constructs that can be used to describe the key business purpose of a solution. These patterns describe the objectives of the solution, the high-level participants that interact in the solution and the nature of the interactions between the participants ... Each pattern is self-contained. The scope of each pattern embraces the minimum end-to-end flows necessary to implement an automated business process.

who work solely with information. A more meaningful description would perhaps be 'symbol worker'. Everybody, even a roadsweeper, uses knowledge in his job.

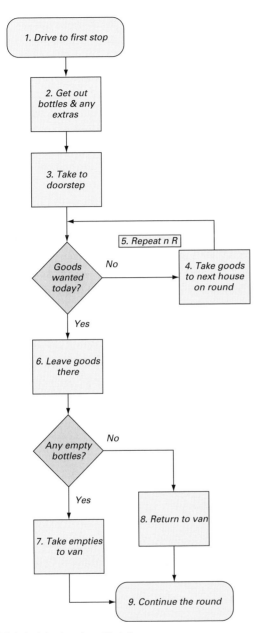

Figure 3.9 A decision tree for milk delivery

The description goes on to make the point that patterns should be able to interact with other patterns, through one or more 'integration points'. These might include file transfer, message transfer, a common database, common access point or a common workflow. Some commentators refer to such interlinked catalogues of patterns as providing a pattern language, a term derived from architectural practice.

Patterns for software development are also growing in popularity. Done well, they reflect the experience, knowledge and insight of developers who have used them successfully in their own work. As well as being reusable in themselves, they provide reusable knowledge, mainly in the form of a common vocabulary.

On their own, business process patterns cannot guarantee success. Any attempt to treat them as plug-and-go business frameworks courts failure. They need modifying to the circumstances of the adopting organization, its people, resources, strategy and so on.

Providing that need is recognized, software patterns can speed development, impart consistency and improve software quality.

Programs, programs everywhere

We said earlier that processes have always been all around us. In recent years, with the rise of the personal computer and the World Wide Web, computer-based processes have added greatly to their number. It hardly needs saying that the programs on our PCs are all managing processes. Indeed, a program is a description of a process. Our computers are also constantly managing themselves through inbuilt processes, mainly under the control of the operating system. Automatically updating email virus definitions is an example.

What is less obvious to the ordinary user is how many other little programs and processes he has dealings with when using his PC. They go by various names – macro, script, wizard, applet, daemon and agent are just a few. These mini-programs perform some small task, often needing action by the user. Examples include subscribing to an Internet service, setting the options on a newly installed game or, indeed, installing the game. These all involve computer-managed processes. There is branching logic behind each of them, the path they follow being determined completely or mainly by what the user inputs.

If you have ever written a macro for your word processor or spreadsheet program, you have even designed that process. A simple macro uses only straight-line or unbranched logic – click on the new icon on the toolbar and the document adopts landscape format, for example – but it is still a process. Make it more fancy, so it waits part-way through for you to tell it what to do next, and it becomes branched. You're a programmer!

Branching logic (figure 3.10) is not confined to the computers you can see. Whenever you do battle with a telephone call management system or retrieve a message from your cellphone, you are in the embrace of conditional logic. It also comes into play when you draw cash at the hole in the wall machine and specify a receipt. And so on, almost ad infinitum.

Computerized logic manages actions on an industrial scale as well. Take printing presses. Large modern printing machines are usually computer-controlled and often linked electronically. This allows a degree of automation of page size, volumes,

Figure 3.10 Branching logic everywhere

speed, paper type and other variables. It also allows integrated job forecasting and aids load sharing.

The presses can work out from the jobs loaded on to them how much ink of which kind they need. They then monitor the rate of use of those inks. Once the amount of any ink reaches a pre-set reorder level, the machine automatically sends a supply request. This goes to the ink supplier to place the order and also triggers a standard purchasing procedure. These two procedures then negotiate with each other to ensure that the new ink arrives on time at the agreed price. Software agents able to move around networks do this. There is no human intervention and no one knows it is taking place. There is after-the-fact management information available to review when needed or should exceptions arise.

Such self-managing processes, while producing clear business benefits, need careful planning. Something on a grander scale probably contributed to the 1987 'Black Monday' crash on the New York Stock Market. The Dow–Jones Industrial Average of share prices fell 22 per cent in one day: this was nearly double the fall in the 1929 stock market crash that set off the Great Depression.

Some observers blame the 1987 event in part on the way different stockbrokers' programmed trading systems triggered one other. There was no overriding control and each system unrestrainedly magnified the outputs of other trading systems. The result was a dramatic and expensive demonstration of the effects of positive feedback.

In chapter 4, we look at software that should have a more beneficial effect on your organization's processes.

Types of process management software

We will return later to the subject of embedded systems for process control. Here, we start by looking at 'ready-wrapped' process management products that your organization can buy and use. There are four main types.

Electronic forms

The simplest way of process managing data is to employ electronic forms (e-forms). In the main, these are cheap and easy to use. They include interactive design tools to allow the layout of paper forms to be imitated on the computer screen. Forms go to other users by electronic mail and arrive in the recipients' in-basket as mail items. An e-form can access external databases and can also enter data to ('populate') them from the contents of specific fields. Manual entry of data to e-forms is also possible, as is adding electronic signatures.

There was once a thriving market for e-forms but they have largely become commodified by the makers of PC software. Of the few specialist suppliers remaining, Accelio (formerly Jetform) and Cardiff (bought by Verity in March 2004) are among the best known.

These days, most e-forms suppliers offer products able to use XML, HTML, Adobe PDF formats or all three.[5] Increasingly, e-forms products conform to Microsoft's InfoPath specification, which uses XML to create forms consistent with its Office 2003 range of products.

Whatever the foundation for the software, the weakness with e-forms has always been that users cannot control the route a form takes once it has been sent. Also, the tools given to users to monitor progress and workloads are often meagre. Forms-handling products cannot be upgraded to other kinds of process management software, such as those we describe below.

Despite these limitations as self-contained products, e-forms are an important ingredient of most process management systems. They form part of the user interface for most workflow and electronic document management products. These usually include forms design tools.

Electronic document management (EDM)

These systems manage sets of electronic documents or their constituents. They let users index, store, view, retrieve, reuse and transfer those files. Files usually contain

---

5 See the glossary if you are unsure what any of these are.

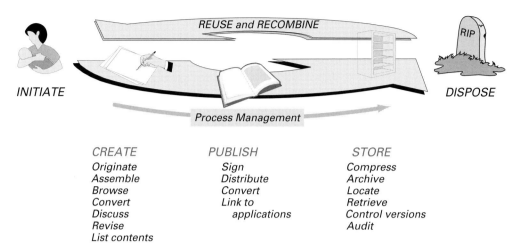

CREATE	PUBLISH	STORE
Originate	Sign	Compress
Assemble	Distribute	Archive
Browse	Convert	Locate
Convert	Link to	Retrieve
Discuss	applications	Control versions
Revise		Audit
List contents		

Figure 3.11 Life-cycle of the electronic document

free-form ('unstructured') data, such as word processor files. These can be searched ('free-text retrieval'). EDM systems also store and manage scanned images of paper documents, spreadsheets and engineering drawings. The constituent parts of files can be separated out and reused in different formats and combinations ('repurposed')(figure 3.11).

Most document management software applies central control to the master copies of files. Authorized users can deposit and retrieve these, a routine called check in and check out. This prevents confusion over which version of a document is definitive. The software keeps track of multiple copies of these definitive files, identifying the 'children' of master versions.

Users can also reassemble and recycle complete or partial stored files. (Products that do not permit this do not warrant being called document management software, in our view.)

EDM systems often link to transaction or line-of-business application programs, such as for insurance claims processing.

Underlying all these capabilities are integral process management tools, allowing files to be managed around a network. With some of the more basic EDM products, this amounts to little more than despatching a file to a pre-set distribution list. Users can keep track of the file's progress and can possibly do some limited conditional routing. What they cannot do, which more sophisticated products allow, is control the file's progress once it has been despatched. In these cheaper EDM products, monitoring of overall progress and workloads is also inferior. Good document management systems offer more complete control everywhere on the network.

Where more complex tasks arise, such as controlling other processes and application programs, a full-scale workflow product is used as well or instead. Indeed, if an EDM product can do this on its own, we view it as a workflow automation product. Products and product categories evolve. For simplicity's sake, we include within this class of product some related systems:

- Document image processing (DIP) is, as its name suggests, designed to manage files that wholly or mainly consist of images of paper documents. These often include workflow management tools but mainly for routing files.
- Engineering drawing management (also EDM) works mainly with files from computer-aided design (CAD) systems and with scans of paper drawings
- Electronic content management (ECM) handles the files that go to make up a World Wide Web site or page. These systems are optimized to manage rapidly changing material
- Electronic records management (ERM) is a term found mainly in governmental, military and other non-profit organizations. Some treat it as a synonym for EDM, as defined above. Others extend the idea to also include any associated manual procedures, training, budgeting and management policies.

Workflow automation

'Workflow', as it is often called, is the use of software to model human work processes on a computer. It passes electronic files, information or tasks from one human participant to another for action. In doing so, it enforces compliance with defined sequences and procedures, embodied in rules.

This software helps organizations conduct their processes speedily, accurately and consistently. Nothing gets left out, nothing gets skimped and nothing gets done twice. And, because managing those processes is a part of running them, everything is controllable, measurable and auditable.

A frequently used analogy is with a relay race. Responsibility for dealing with a work object such as an electronic file passes from person to person until the process is complete. It is a sequence of mutually dependent activities.

Anybody who has watched a sprint relay team in action can see in just how many ways this can go wrong. For example, when the first athlete gets to the changeover place, he might discover his team-mate has come out the blocks too soon. If he can't catch the second runner in the appointed distance, the result is disqualification. Alternatively, if the second man starts too late, he has to jog until the first runner catches up with him. The result is a slow time. A third possibility is that, between them, the runners fumble or even drop the baton. A slow time and disqualification are the respective outcomes here, too. Workflow automation promises fast circuits of the track with no baton-dropping and no disqualifications.

Useful as it is, the analogy should not be pressed too far. Relay runners go round the track in just one lane, for example, and in one direction. When a runner receives a baton, he does not inspect it and, based on that scrutiny, decide to run across the infield or out of the stadium. Nor does he break the baton in two and give one piece to a fifth team member to carry, reassembling the pieces before the finish line. Workflow software permits the equivalent of all these actions and more. A successfully run relay race uses unbranching logic; a successfully run business process usually uses conditional logic.

Workflow automation is not new. Large information-handling organizations such as banks and insurance companies have been using it for at least twenty years. It has allowed them to manage repetitive operations such as policy issue and claims handling that would otherwise be expensively labour-intensive. These organizations use software optimized to handle large volumes of short-cycle tasks, involving relatively simple processes. This software often links to a document management system, to automate correspondence handling. These days, there are also links to call-centres.

These 'paper factory' centralized systems are no longer representative of most workflow software installations. Newer workflow products embody a more variable model of office life, with parallel workflows, in-flight variation and user design of the workflow paths and actions. This kind of workflow is increasingly used as a support service for professional staff.

The benefits of workflow automation can apply not just to routine work, as the case studies throughout this book show. At its furthest extreme, some processes can take years to unfold and might arise only once or twice a decade. Building an oil rig and developing a new drug are two examples. In both, saving effort and elapsed time are important business drivers. In pharmaceutical companies, especially, there are also legal and regulatory demands to meet when recording actions and decisions.

Whatever their purpose, workflow installations depend on the initial creation of a detailed and dynamic model of the processes to be automated. We described this stage earlier in the chapter.

Once agreed and completed, the model is stored on the workflow server. This then controls the processing of each case ('instance') of the procedure). At each running, the software applies the rules and conditions contained in the model.

The workflow server works into a database system. This is either built in or, more often, from one of the major database software makers, such as Oracle or IBM.

Together, the server (or 'engine') and the database provide services to each individual user's client machine. Those services include notifying new tasks or instances, managing work queues and triggering associated programs. These other programs can be on the client machine, such as a word processor, or at the back end, such as a central accounts package. Processes can be interrupted or changed while in progress ('in-flight' changes).

Most modern workflow software will allow a designer to create process steps that are left to human discretion. Users are then free to deal with unexpected events or those with too many variables to pre-program. The subsequent or downstream process will vary according to the results of their research or decisions.

A business process can be regarded as made up of three components – actions, the sequence in which they are taken and their timing. Good workflow software accommodates *ad hoc* changes to any of these and manages what results.

The import and export of data to other application programs is a standard feature. It usually takes place dynamically, while these programs are running. The workflow system can also invoke those programs, starting them as needed to perform some task or to retrieve or send some data. Where there is no subsequent feedback to the workflow system, this is sometimes given the disparaging name of 'fire and forget'.

Modern workflow products can manage processes that span the organization. Some can also manage processes that run between organizations. Unless those organizations are using the same software, the ability to interwork will depend on standards.

Fortunately, workflow products are becoming increasingly open in nature. Closed, proprietary systems allow integration and interworking only if an entire organization's supply chain uses only that supplier's products. This is an increasingly rare occurrence and such products are dying out. Open, standards-based products lay a foundation for greater inclusiveness and adaptability.

Embedded and infrastructural workflow

As we have described it so far, workflow software has been an explicit, visible application for the user. In other words, he is aware of using the software. He will engage with it either through its maker's proprietary client or through a standard client, such as Microsoft Outlook or a Web browser.

This is becoming a minority situation. Software makers are incorporating workflow products – their own or other makers' – within broader offerings. Systems integrators do this, too. Pre-press systems in the publishing industry, for instance, often have workflow capabilities built in, as do full-service document management products.

Using software products this way is termed embedding. One result of it is that the user is usually unaware of the workflow supplier's identity. Often, he is unaware of using workflow software at all. The workflow product has, so to speak, been distanced from the end-user.

This concealment is complete when the software containing the workflow engine provides services to other software and systems. By this stage, the workflow software has become middleware, part of the systems plumbing for the organization.

Modern process management software can thus be any or all of three things – a stand-alone product, embedded software or middleware. In this last guise, it becomes an infrastructure product, aimed at process-enabling the enterprise. Typical of the breed is the Fiorano Business Integration Suite, as used by POSCO (see the case study in chapter 7). Fiorano refer to it as an 'enterprise process backbone', which is an apt description.

Workflow and groupware

Groupware is software for enabling and aiding collaborative human working. Email is also intended for person-to-person (or person-to-group) working and can thus be counted as groupware. This kind of software became popular largely through the wide success of Lotus Notes and its later imitator, Microsoft Exchange. Groupware makes interaction between human beings the primary concern of a computer system.

Most other application software manages interactions either with data or with processes, the latter typically in the shape of other programs. There are thus three kinds of interacting entity – people, data and processes.

Table 3.2 shows some examples of these different interactions. Note that not all these activities are easily placed in just one pigeonhole. The selections we have made are by way of illustration. The categories refer to the main actors in the system.[6]

The person-to-person activities (top left cell) are managed by groupware. Those at the top right are taking place between a person and process management system, such as workflow software. The other cells in the right-hand column are those typical of behind-the-scenes process management systems, such as the Fiorano process backbone product.

Table 3.2 Interactions with data and processes

To:	Person	Data	Process
From: Person	Electronic mailing	Electronic form entry	Using a project
	Videoconferencing	Scheduler entry	management system
	Cellphone texting	Word processing	Making an ATM withdrawal
			On-line shopping
Data	Web page display	Web cookies	Email encryption
	Fax receipt	(sending and detecting)	Stock management
	Television viewing	Database replication	system entry
		Credit checking	
Process	Getting a workflow	Search engine 'crawling'	Inter-server workflow
	notification	Email filtering	(via enterprise backbone)
	Calendar reminder		Online 'trojans'
	System error message		

6 All these actions are mediated by computer, so could pedantically be classed as only process or data handling but that would be to miss the point.

Data-gathering is taking place in the top centre cell, data distribution centre left.

These distinctions are important when specifying and designing a system. For example, a system optimized for data-to-data communication would not only be inadequate for human-to-human interaction, it would most likely impede it.

Some commentators mistakenly describe workflow software as aiding 'people-to-people processes'. Such processes exist but not once you interpose a computer between those people. The interaction necessarily is then between the person and data or a process. The upshot is that, with the exception mainly of Lotus Notes, process management systems are not groupware nor is the reverse true. Even then, Lotus Notes' workflow capabilities are limited compared with specialized process management products.

BPM

BPM software builds upon the foundations laid by workflow software. It has four main distinctive technical features:

- It not only links with but can also integrate other application programs. At the computer programming level, BPM is the universal connector. It is the software equivalent of those handy devices that you buy at airports for running electrical equipment while abroad.
- It can manage those programs, giving central or local control of entire processes, from end to end. (Think of BPM as also standing for business *programs* management.)
- It gives users a graphical real-time view of what is happening to all the processes and applications it connects with. In suppliers' parlance, it provides BI or BAM.
- It lets authorized users easily change those processes or add new ones, though a separable 'logic layer'. This is sometimes called a business rules engine.[7] It can make these changes either before an instance of a process runs or while it is in progress.

Modern business processes are now invoked through many different channels, including the Internet, call-centres and ATMs. The resulting range and complexity of processes makes an enterprise-wide approach to managing them the logical next step.

At the moment, each different application program is in charge of its own set of processes. To drive these, it tries to exercise dominance over adjacent programs. This may sound overly dramatic but is the literal truth. Applications programs compete with each other for computing resources, network capacity and the attention of other programs.

---

7 Although called an engine, it is a piece of software. Most rules engines work through a decision tree, of the kind we described earlier for the milkman. They use it to arrive at the instructions they will give the process management software.

Table 3.3 Process management software compared

	Ability	EDM	Workflow	BPM
Data handling	Makes electronic versions of paper documents?	Yes	Uses EDM for this	Uses EDM for this
	Handles all kinds of digital document (e.g. word processor files)?	Yes	Yes	Yes
	Indexes and searches digital documents?	Yes	Uses EDM for this	Uses EDM for this
	Allows the reuse of documents, and the disassembly and recombination of their components?	Yes	Uses EDM for this	Uses EDM for this
	Allows users exclusive access to master documents (check in/out)?	Yes	Uses EDM for this	Uses EDM for this
	Links digital documents with other data?	Yes	Yes	Yes
	Exchanges data with other software?	Yes	Yes	Yes
Routing	Routes digital documents around a network?	Yes	Yes	Yes
	Manages the speed and direction of the route?	Sometimes	Yes	Yes
	Sets up rules for managing routes, timing, actors, etc.?	Sometimes	Yes	Yes
	Runs processes outside the organization?	No	Sometimes	Yes
Process management	Modifies processes on the fly (that is, without having to stop)?	No or seldom	Yes	Yes
	Manages human activities?	Yes	Yes	Yes
	Invokes the running of other software	No	Yes	Also integrates and interacts with other software (orchestration)?
	Manages an organization's other software	No	No	Yes
	Uses, and is used by, Web services?	No	No	Yes
	Offers goal-directed processing?	No	No	Sometimes
	Offers predictive processing?	No	No	Sometimes
Process design	Graphically models processes and creates the rules to run them?	Sometimes	Yes	Yes
	Simulates and optimizes processes?	No	Sometimes	Yes
	Manages the rules from application programs separately?	No	No	Yes
	Allows analysts and users to add or modify rules?	Sometimes	Sometimes	Yes

Table 3.3 (*cont.*)

	Ability	EDM	Workflow	BPM
Reporting	Provides after-the-fact reports and audit trails?	Sometimes	Yes	Yes
	Provides real-time reports on progress and performance?	No	Yes	Yes. Also provides real-time reports on progress in all connected application programs (BAM)
Product design	Open programming interfaces?	Usually	Yes	Yes
	Standards-based?	Sometimes	Yes, usually	Yes, always

BPM software takes the control of business processes away from the individual application programs. Instead of contending with one another, these programs become subordinate to the BPM software and the business rules it contains. The logic layer controls their execution of the processes. It delegates tasks or activities to the individual application programs according to their strengths and the needs of the moment.

To do this well, BPM needs to support all the attributes of a business process, which we described above. For example, it needs to be able to manage:

- application programs in parallel as well as in series
- people-intensive applications
- processes inside and outside the organization
- both continuous and discrete processes, and allow them to change with time.

In effect, the BPM system acts as a superstructure that brings together these activities into single interconnected entity. This can be understood, managed and changed as though it were one system.

Process management systems compared

BPM software has its heritage in workflow automation, and the dividing line between them is sometimes hard to see. The minutiae of this debate matter deeply to software makers and industry analysts but less so to users. They are sensibly more concerned with what the software can do than with marketing labels. Table 3.3 offers some thoughts on those differences. The main distinguishing features are those we have listed above.

We have tried to be fair in compiling this table. It is easy to exaggerate distinctions in such exercises, as many suppliers' literature makes clear. We have tried not to fall into that trap or to create straw men in our summary of the abilities of the older generation of products. These are still valuable and, indeed, may be a more cost-effective option in some situations. We have, though, excluded electronic forms because of their limited capabilities.

In Chapter 4, we look at the role BPM plays in relation to other corporate systems, computer-based or otherwise.

Case Study 2 DVLA

The Driver and Vehicle Licensing Agency (DVLA) is the licensing authority for motor vehicles and drivers for the UK. Its work consumes and produces enormous quantities of paper. The Agency had an ageing DIP system but no systematic computer-based control of processes throughout the organization. Its responses to enquiries were consequently slow and error-prone. DVLA decided to introduce BPM to improve its performance.

One of the most complex areas of the Agency's work is where a driver has some form of medical condition or history that could affect his ability to drive. The case study concentrates on the introduction and use of the BPM system in the group that deals with these cases. Called CASP (Casework and Specialist Processes), it needed a new system infrastructure before becoming fully productive. It also had to link to a large amount of historical data, much of it on microfilm.

The design team used RAD and joint application development (JAD) techniques, in which users, system developers and equipment suppliers all took part. As a result, few changes were needed after the system's introduction. Human and cultural problems appear to have been minor. Productivity, case processing speed and customer satisfaction have all risen since introducing CASP. This is a large and pervasive system. Its expanding use will help turn the whole of DVLA into a speedily reactive and efficient organization.

Industry/Sector	Central government	**Location(s)**	Swansea, UK, plus 40 UK local offices
Annual turnover/ income	£312 million (2002–3)	**Number of employees**	7,600
Type of system	Workflow	**Supplier and product**	TIBCO Staffware Process Suite
Number of users	300 users in DMG and more than 1,370 across DVLA; these numbers are set to rise	**Time to complete**	Phased implementation over 3 years
Business objectives	• Improve overall business efficiency • Improve customer satisfaction and service • Reduce dependence on paper • Give greater visibility of exceptions processing between departments • Provide better and more up-to-date management information		
Quantitative results	• 24 per cent increase in business transactions and cases handled • Processing more than 3,000 cases daily, up from 1,800–2,000 cases managed by earlier system • Handling times of transactions cut by days		

Qualitative results	• The system is more reliable, robust and secure • Staff satisfaction has increased significantly through better visibility of cases • Releasing office space has allowed two new teams to move into space in the operational area previously occupied by filing cabinets • The customer experience has improved dramatically

Business background

The DVLA is part of the British government's Department for Transport. The Agency is based in Swansea, in south Wales, and has forty regional offices. Its responsibilities include:

- keeping records of licensed drivers and vehicle keepers
- issuing and, where needed, withdrawing licences to drivers
- issuing registration documents and annual vehicle licences
- collecting and enforcing vehicle excise duty ('road tax')
- issuing vehicle registration marks ('number plates').

The DVLA keeps registers of over 30 million vehicles and more than 40 million drivers. In doing so, it processes over 91 million forms and carries out 50 million transactions a year. It can handle most of these quickly but exceptions occur.

Those exceptions about a driver's medical condition go to the Drivers Medical Group (DMG) within DVLA. This group deals with licensing requests from people who have a medical condition that could affect their ability to drive or who have had a conviction for high-risk drinking and driving.

The procedures and checks involved with these higher-risk drivers are more extensive than normal. As a result, the process takes longer and involves more stages and paperwork. Most will need extra information from the driver and from the medical profession and possibly also data from the courts and police.

There are 300,000 of these cases a year. The DMG relied on a DIP system to help handle the resulting paperwork. This had been in successful service for nearly fourteen years but was obsolete. There was no support available for the software and the system was becoming unreliable. It was also unable to share cases between departments electronically. Archiving for other groups was on 16mm microfilm, consisting of over 1 billion images. These groups were Vehicle Customer Services (VCS), Driver Customer Services (DCS) and Customer Enquiry Vehicle (CEV).

Anita Evans is one of the DVLA's Project and Programme Support and Assurance Managers, with responsibility for the project. She says: 'If we had a driver whose case history resided within DMG but needed to be shared with another department, it had to be specially scanned. For this, you had to pack up a box of paper files and send them through the internal postal system to another building two miles from campus.'

The problem did not stop there. 'Every department worked in a silo. Customers would write a letter to one department to get information and then have to write a second letter to get data from another department. Caseworkers would have to put customers on hold to dig out information from another system, or call another colleague. Often people would get sick of waiting, hang up and call back.'

In 1999, the DVLA began a project to handle the information for DMG and other groups by computer. Its three objectives were to:

- process more cases, leading to increased customer satisfaction
- give greater visibility between departments
- provide more – and more current – management information.

The Agency decided that this new system should also be able to:

- handle errors and problems speedily and easily
- re-route cases
- help with producing letters
- link to the microfilm records
- link to other internal systems and databases.

The result was the CASP system.

System description

The CASP system is based on TIBCO Staffware BPM software. It orchestrates the casework, progressing each case along defined work paths and following pre-determined parameters and rules. The system oversees each case, giving management control, and provides continuous details of work efficiency. There are automatic alerts whenever system parameters are exceeded, or where rules and tasks are overlooked. The new system comprises a 6 terabyte distributed storage system, linked to six Unix servers. These run Oracle databases, a Tower imaging system and the TIBCO Staffware Process Suite. There is a link to an automated microfilm retrieval system. Some servers running Microsoft Windows provide support functions, such as improved letter creation.

Presently, there are more than 300 users in the DMG and over 1,370 users throughout DVLA.

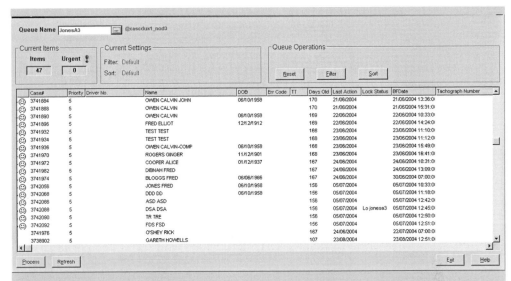

Figure C2.1 Sample user's screen for CASP

Implementation experience

Where and when did the project or system originate?	The CASP project began in 1999; planning and infrastructure changes have meant that the TIBCO system has been live since February 2002
How long did implementation take?	About 3 years, taking into account infrastructure changes, design, build and phased implementation
Who did the implementing (own staff, contractors, consultants, etc.)?	Several companies helped with the project, including:
	• Anacomp, which was responsible for technical consultancy and back file conversion
	• EDS, which implemented the system for the first business area
	• Fujitsu, which completed the installation for the rest of the organization; it also developed the link into the vehicle casework section
How much bespoke development was there?	A significant amount of development was needed to achieve a single, simple user interface; using Visual Basic, this linked Staffware, the Tower image viewer and the Oracle databases
Were there any special infrastructure needs?	The whole infrastructure needed some development, mainly to satisfy an Agency-wide need for resilience to failure

What were the most significant implementation issues and how were they dealt with?	Converting over 20 years' worth of back files; Kodak built a special system for this.
Who is responsible for the system overall?	Fujitsu, as sub-contractor to IBM under DVLA's partnering contract with IBM
How was and is training handled?	In-house and by each department; selected users got involved in user acceptance testing and training, and were then able to train colleagues
What was and is done to encourage use?	Its use is compulsory
What lessons were learned?	• To ensure the infrastructure for design is in place before looking into solutions • Tighter controls on the scope for a phase 1 launch would have achieved faster implementation, with 'nice to have' features built a phase later

Anacomp and EDS led the analysis and design of the system, using RAD and JAD techniques. As part of this, users, system developers and equipment suppliers took part in shared workshops. Any necessary changes were made before the system was finalised, so hardly any changes were needed after the systems' introduction.

Back file conversion was a major task, converting over twenty years' worth of data (24 million images) to a different electronic format. Kodak developed a system to translate microfiche records into electronic form. A new technical infrastructure was necessary as well. Chris Haden is Managing Director of Anacomp UK, which provided technical consultancy during the system's introduction: 'The new casework system required a radical change of architecture to support it, with the existing drivers medical system infrastructure creaking under the weight of work.'

While the new infrastructure was being installed and tested, Staffware and EDS began workshops with DVLA to define the business needs for the CASP project. This looked in particular at the work of the people who dealt with customers, who would have first-hand knowledge of where processes needed improvement. The workshops took the form of JAD sessions. These involved users, system architects, Agency operational staff and technical support personnel. Progressive implementation took place across five operational areas. Managing users' expectations proved an issue. As Anita Evans says: 'To deliver a workable product, the system had to go live without some of the system enhancements quoted in the requirements. For example, we worked with local printing for the standard letter product until the new print server could be delivered and central printing could be delivered.' (Central printing saved twelve full-time equivalent job positions.)

The DVLA began installing the new system in two customer service departments – DCS and CED. The DCS system began working in February 2002 and the CED system in May 2003. The DMG system was the second-to-last phase of the project. This finished in September 2003 after a short delay while project management was transferred from EDS to IBM and Fujitsu.

The DMG system is the most complex. Fewer than 5 per cent of applications in the DCS and CED sections need casework. DMG, by contrast, is all casework. Designing and introducing this was the largest phase of the project.

The main lessons learned were:

- To ensure the infrastructure for design was in place before looking into solutions.
- To manage user needs better. This would have meant less reliance on bespoke programs. The system focused on users rather than specific organizational needs

Benefits and user reaction

What has been the reaction of managers and staff?	Initially, huge reluctance by users to changing their working practices; overcome with the aid of a familiar-looking computer interface
What has been the reaction of customers or trading partners?	Customers are much more satisfied, particularly with the call-centre; they can now get answers in one call
What has been the overall cost of the system?	Not available
What have been the main process benefits?	<ul><li>More cases processed with no increase in staff</li><li>Customer correspondence has been streamlined and standardised</li><li>The 'silo' attitude has been overcome – data is now shared electronically across departments; DVLA can now offer customers a true 'one-stop shop'</li></ul>
What have been the main effects on operating style and methods?	<ul><li>The operating style and methods are now much slicker, with cases being handled on just one telephone call</li><li>Data protection has been improved</li></ul>

Before introducing the CASP system, the DMG could process between 1,800 and 2,000 cases a day. It can now handle over 3,000 cases daily. The Group also has a single, coherent view of the Agency. Cases can go between it and other departments electronically, rather than having to be transported physically. Auditing and tracing of cases have improved and DMG alone cut its backlog of 100,000 cases to 60,000 in six months.

Customer service has been boosted. Now, caseworkers and call-centre agents can access all details on a particular case with a click of a button. There no longer any need to make calls between different departments to try to get the full facts of case. This was previously a major cause of customer dissatisfaction. The previous system was painfully slow and caused call times to average 6–8 minutes instead of the DVLA target of 2 minutes.

Data protection restrictions were increased, to ensure that information was delivered only to the people allowed access to it. For example, a call-centre operator is not allowed to see confidential medical details. Anita Evans is convinced of the ability of process management to transform business practices, increase visibility and provide 'joined-up' interoperability: 'Managing the caseload which millions of drivers and vehicles create is a major task. With the new casework system, the job is far more manageable. There is greater transparency, better sharing of information and accelerated case processing to provide customers with a much higher level of service than ever before.'

She continues: 'We are seeing call numbers go down but average call times go up. Calls are taking much longer than before because the customers are asking more questions and getting them all answered in one single phone call. Customers used to get sick of waiting for an answer, hang up and call back. Now, the calls may be longer but the customer has a far better experience.' 'Correspondence has also been streamlined. Customers now don't get several letters from different departments asking for information; they just get the one, reducing any customer confusion. Importantly, we have broken down the information silo outlook – data is now shared across departments and we can offer the customer a true 'one-stop shop'.'

Evans says that upgrading to new equipment has also provided other benefits: 'The old system crashed often because of its age. It simply couldn't handle the workload it was subjected to, which meant a lot of downtime and productivity dips. We needed a sustainable, supportable platform for case management and now that's what we've got.'

The DVLA is also saving money on employee costs. Although no posts were lost through this initiative, it is now not necessary to replace those leaving, such as through retirement. It has also been able to reassign staff to other areas of the organisation that have a pressing need for personnel. Despite a 24 per cent increase in medical queries in five years, DMG has not had to employ more staff.

Rather than 2 weeks, system training now takes only 2 days. This has slashed the time needed to get employees familiar with the system and has reducing training costs significantly. The older systems were also expensive to support, another cost that has been reduced substantially.

Staff, including managers, were initially greatly reluctant to change their working practices. DVLA eased this problem by ensuring that the look and feel of the new system was similar to what users already had.

Other projects running in parallel affected implementation. If the DVLA were to undertake this project again, it says it would ensure that the infrastructure was fully in place before developing a BPM system. The need to avoid a single point of system failure had significant impact on the CASP implementation.

These and other external factors caused the long implementation time, not difficulties with the project itself. The business had been struggling with a failing system for so long that any delivery of a more robust system was accepted with open arms. As Anita Evans says: 'We were working to a live registry of 102,000 at the time of "go live". The registry today is close to 40,000 after just one year of working with CASP.'

The system has proved flexible and has accommodated changes to requirements with ease. For example, the DCS needed a system for handling exceptions in processing applications for digital ('smart') tachographs. This was added while CASP ran uninterrupted.

The future

With the system now bedded in at the DMG, DVLA's VCS department is entering the user acceptance-testing phase. This is planned for April 2005, when another 1,205 users will go on line. The Vehicles Group (VEG) is the largest operations group within the Agency. The final piece of the CASP project will be when the CEV section adopts the system.

Future intentions are confidential at this stage, but design teams are carrying out feasibility studies to ensure that the BPM system is put to best use.

4   The systems of an organization

The enterprise nervous system

There are many types of process in any organization, not all of them known about or acknowledged. They include:

- Work processes, which are the usual targets of process improvement programmes
- Management processes, which are not always thought of as processes
- Change processes, which are simply not always thought about
- Learning processes, which mainly educators and knowledge management consultants think about
- Budgeting processes, which are thought about but are mainly resented.

These processes and many others contribute to the organization's existence and help it perform its allotted tasks. They can be likened to the human body's nervous system, which enables it to respond to changes in its external and internal environments.

The nervous system consists mainly of cells called neurons, of which there are about 100 billion in the brain alone.[1] Neurons fall into three general types:

- sensory neurons, which relay information from the senses
- motor neurons, which carry impulses to muscles and glands
- interneurons, which transmit impulses between the other two kinds of neuron.

Some parts of the system are under deliberate control. In a healthy person, for example, speaking and moving one's limbs are conscious actions. These are managed by the somatic nervous system (from the Greek, *soma*, meaning body). Other activities, such as digesting food and sweating, are involuntary. These are under the

---

1 If you compare a neuron to a transistor, this is roughly equivalent to having 2,500 Intel Pentium 4 processors in your head – but without the external power supply or cooling problems. Also, transistors can't move around or swap connections; neurons can and do and under their own volition. The brain is self-healing and self-changing, even into old age.

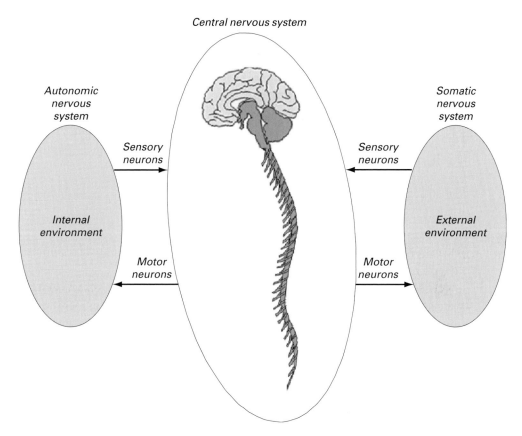

Figure 4.1 Human nervous system

respective control of the autonomic nervous system. (Autonomic means self-governing: more Greek.)

These two parts form the peripheral nervous system. This connects with the central nervous system, made up of the brain and the spinal cord. The functional diagram in figure 4.1 shows these in outline.

The nervous system is often used in analogy when thinking about organizations and their computer systems. Some software suppliers, for instance, refer to their infrastructural products as 'enterprise backbone systems' (EBS).[2] These connect multiple and diverse application programs, rapidly exchanging data among them and coordinating their actions. As we have seen, this is also part of what BPM software does. Where one kind of software stops and the other starts is a matter for

---

2 Data communications companies by contrast use 'enterprise backbone' to mean a high-speed and large-capacity data network that serves the whole organization. Smaller, local networks connect to this.

> **Box 4.1** IBM's view of autonomic computer systems
>
> They:
> 1. Possess a sense of self
> 2. Adapt to changes in their environment
> 3. Strive to improve its performance
> 4. Heal when damaged
> 5. Defend themselves against attackers
> 6. Exchange resources with unfamiliar systems
> 7. Communicate through open standards
> 8. Anticipate users' actions.

debate. Like so many other debates in computing, this one is coloured by commercial self-interest.

Another current term is 'autonomic computing'. This describes systems that can manage themselves, typically with high-level guidance from humans. Most of the large computer suppliers, especially IBM, Hewlett-Packard and Sun, are investigating the possibilities of these. Box 4.1, shows IBM's list of ideal characteristics for such a system. The similarity with living organisms is both clear and deliberate.

Cybernetics and management

A close cousin of the nervous system analogy is the cybernetic view of organizations. This sees managers' jobs as being mainly or solely to regulate, govern or control operations.[3] Probably the most detailed exposition of the similarity occurs in the writings of Stafford Beer. The theme of these was something he called the Viable System Model. Beer used the word 'viable' – able to live a separate existence – because he saw organizations as living things. Like all organisms, they must learn and adapt to survive. A system is viable if it is capable over the long term of responding to environmental changes, particularly those not foreseen when the system was developed or designed. Prolonged equilibrium, on the other hand, is a precursor to death.[4]

---

3 The mathematician Norbert Wiener gave the word 'cybernetics' its modern meaning in 1948. He used the word in the title of his first book on the subject, *Cybernetics, or Control and Communication in the Animal and the Machine*, which helpfully explains the term. It derives from the Greek, *gubernatos*, meaning steersman, and shares a linguistic root with the word 'governor'. There is an earlier French use, from 1854, to mean the science of civil government.

4 We borrowed and adapted this sentence from the book by Richard Pascale, Mark Millemann and Linda Gioja: *Surfing the Edge of Chaos: The Laws of Nature and the New Laws of Business*.

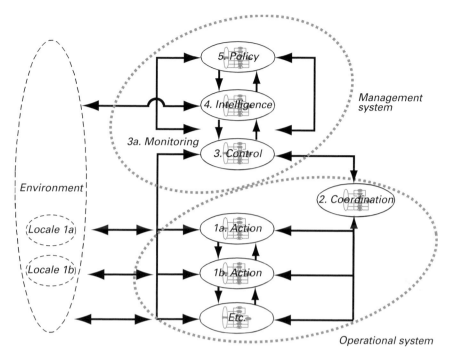

Figure 4.2 The Viable System Model

Survivability is the watchword. Just because something is alive does not make it viable. It has to be able to continue living. Many existing organizations or units that look healthy are in fact dying or are unfitted to survive. As with the human body, the pathology of organizational diseases includes many that hide their symptoms from superficial examination.

Figure 4.2 is a diagram of Beer's model, adapted slightly to make it easier to understand.

There are six main entities depicted, the largest being the *environment* into which the system works. The other five are sub-systems of the organism that exists within that environment. These sub-systems are essential to its independent survival.

1. **Actions**, the primary operations and activities of the unit under study, its *raison d'être*. Each operation links to a locale, that part of the environment immediately relevant to the particular action (shown as action 1a linking to locale 1a, and so on). In a sales function, for instance, the locale would be its current customers and prospects. (The analogy at this level is with the muscles and major organs, acting under the control of the somatic nervous system.)

2. **Coordination**, which synchronizes and regulates activities and allocates resources. It makes sure the various actions at level 1 do not compete with or overlap one another unhelpfully. (Here, the analogy is with the autonomic nervous system.)

3. **Control**, which is based on the direct monitoring of actions (sub-sub-system 3a) and the inputs from coordination (sub-system 2). It can be exercised through those same routes. This sub-system provides equilibrium, support and audit. (In the body, these are the responsibility of part of the hindbrain.)

4. **Intelligence**, gathering external information and linking it with internal information. Unlike the first three sub-systems, which deal with the immediate and local, this deals with future and global matters. (The analogy is with the activities of the *diencephalon*, which lies above the midbrain.)

5. **Policy**, giving overall direction, including strategic planning. Part of this is monitoring the interchange between sub-systems 3 and 4. (In the human body, policy-making is done by the cerebral cortex, the 'grey matter'.)

Sub-systems 1 and 2 form the operational system and sub-systems 3–5 together make up the management system. Some writers refer to levels 3–5 as the *meta-system* to the production system (that is, they are about and above it).

The model can be scaled up or down freely and contains multiple instances of itself. Each instance is embedded within the next larger version, much like a set of Russian dolls or *matryoshka*. As Beer put it: 'at whatever level of aggregation we start, then the whole model is rewritten in each element of the original model, and so on indefinitely.'[5] To emphasize this, we have drawn a small version of the entire model within each of the five sub-systems.

An important consequence of this replication is that the management functions (3, 4 and 5) take place everywhere the model applies, not just at the 'top' of the firm. Whether an organization chart shows it or not, management is always decentralized to some extent. (See chapter 6 for more on the misleading nature of organization charts.) The management system cannot survive without the operational system. The reverse is not the case; formal management is an optional extra, at least in the short term.

For our purposes, the smallest instance of the model arises with the individual worker. We can then apply and expand it all the way up to the trading chain, embracing the entire organization and its suppliers, trading partners and customers.[6] Table 4.1 shows how the model might apply at either end of that size range.

The number of instances, or embeddings, of the model to consider depends on the size and complexity of the organization. For a large firm, the major sub-systems would be its subsidiaries, operating divisions or functional groups. So, for example,

---

5 This characteristic, otherwise known as self-similarity, is the basis of fractals.

6 The Viable System Model applies down to the level of a single cell in our body and, multiplied up, as broadly as a whole nation. During the early 1970s, Beer devised a system based on the model to help President Salvador Allende manage the Chilean economy. Data was transferred and aggregated almost in real time (there was a one-day lag at most) over a national network of Telex machines linked to two central computers. A bloody *coup d'état* in 1973 brought the project to a sudden end. See Beer's *The Brain of the Firm* for more details.

Table 4.1 Application of the Viable System Model

Sub-system	For the individual worker	For the trading organization
0. **Environment**	Workgroup, supervisor, customers, etc.	Government, economic and political conditions, labour markets, customers, suppliers, competitors, etc.
1. **Actions**	Carrying out allotted or self-instigated tasks	Buying, making, selling, supplying, etc. The productive activities of individual plants or offices
2. **Coordination**	Diary, work records, task lists, informal networks and links	Standards, scheduling, meetings, project management, production control, company newsletter, online discussion groups
3. **Control**	Own judgement, and by supervisor or manager (that is, the next level out)	Operations management (continuous), auditing (periodic)
3a. **Monitoring**	Work records, customer reports, progress against targets, etc.	Management accounts, quality reports
4. **Intelligence**	General reading, conversation, feedback, etc.	Market surveys, analysts' reports, research
5. **Policy**	Personal, group and organizational targets, training, conscience . . .	Strategy-making, long-term planning, these can be carried out in a centralized or dispersed manner

strategy for the whole organization informs and is informed by the planning at divisional level. Within divisions, the next level of attention might be departments. Workgroups would probably come within those and then individual workers. As with so much else in organizations, the choice of systems to be examined and their boundaries is a matter of judgement.

Note that none of the entries in table 4.1 necessarily corresponds to a job description or part of it. At each level, it describes the functions to be performed by that part of the greater system. The entries can be the responsibility of an individual or a group. That said, there is no harm and possibly some good in basing job descriptions on the Viable System Model. It will bring a contextual clarity to the job and its relations with others that is often lacking.

A question of variety

One of the most important ideas put forward by the cybernetics school is the need to be able to handle variety. Simply put, this says that any system or organism must find a way of coping with the variability of the world outside it. It can deal with this in three main ways – dampen down outside variations before dealing with them, try to match them or some combination of the two.

Dampening – or filtering – carries risks, especially if carried out crudely or unintelligently. Important signals get missed or distorted; less relevant signals get overemphasized. The tool or process that is doing the filtering must therefore be sensitive to every change in order to 'know' how to tell the difference between the significant and the trivial. In other words, the sensing apparatus must possess as much variety as the environment it is being asked to monitor. This quality is known as the 'requisite variety'.[7]

The same demand is made of a system or organism that tries the second approach, of trying to match the variability of the world outside. Instead of reducing variety by filtering, it increases its own internal variety to emulate that found in its environment. Although theoretically possible, this is impossible in practice. The range of possible states in the organism is always smaller than can be found in its environment. In other words, it is always less complex and variable than its setting.

To stay alive and to preserve its internal equilibrium (the same thing, really), an organization must be able to deal quickly and accurately with externally imposed variety. Only if it lives can it then set about achieving its purposes.

Dealing with the unruliness of the outside demands suitable information and processes. The watchwords here – for surviving and for prospering – are relevance, completeness, timeliness and frequency.

First, information must be relevant to the need or intent. This means it must be appropriate, accurate, undistorted and understandable. It is no use, say, sending messages in programming code to the designers in a fashion house. It is equally unhelpful to send comprehensible reports based on biased data-gathering.

Actions also must be relevant. If customers are asking for new colours or finishes, for example, only a few of them will be won over by supplying more of the existing types. Henry Ford got away with providing nothing but black but modern customers are more demanding.[8]

---

7 Requisite variety is the ability of a system or part of it to respond to or deal with the unusual. The idea of requisite variety first appeared in Ross Ashby's book, *An Introduction to Cybernetics*. As he explains, variety is a close relation to information. Here is a simple example. Imagine a light sensor able to distinguish only white light. It can deal with only two states – white light 'yes' and white light 'no'. Now shine a red light on it or a green. Will it know what to do? Of course not; it does not have enough variety within itself to cope. Exposing it to more than one kind of light means that it will either not respond at all or else give false readings.

8 James Womack, in his article 'Lean Thinking: A Look Back and a Look Forward', points out that Henry Ford was engaging more in variety *avoidance* than reduction: 'The Model T was not just limited to one color. It was also limited to one specification so that all Model T chassis were essentially identical up through the end of production in 1926 … Indeed, it appears that practically every machine in the Ford Motor Company worked on a single part number, and there were essentially no changeovers.' A production engineer's dream, perhaps, but these days a marketer's nightmare.

Third, information and processes must be timely, which encompasses the qualities of punctuality, speed of delivery and up-to-dateness.

Finally, frequency. If the environment is changing hourly, it is no use reporting changes daily and making course corrections monthly. Conversely, if matters change slowly, more frequent reporting or action is wasteful and can be distracting.

In chapter 2, we discussed the problems besetting the delivery manager of a fictitious retailer, Marks, Lewis and Skinner (ML&S). This unfortunate individual was assailed by difficulties in almost every direction. He had no control over customer shipments from the depots, and got only daily reports of the deliveries that suppliers made direct. To make matters worse, he also got details of customers' online transactions only once a day, and those arrived a day late.

This delivery manager was certainly suffering from a lack of timeliness and frequency in information. What he got was possibly low in relevance as well. These deficiencies made it impossible for him to apply suitable corrective action. The company's problems will be echoed and amplified up and down its supply chain, wreaking havoc among its trading partners.[9]

In Viable Systems Model terms, the monitoring (3a) and control (3) within this retailer are nowhere near good enough to permit appropriate actions (1) or their coordination (2). Unless the intelligence-gathering (4) is similarly sub-standard, this points to bad or absent board-level decision making (5). If the intelligence gathering (4) *is* sub-standard, the IT function is falling down on the job.

ML&S' internal communications and processes look to be badly out of tune with the needs of the market. This organism's sensing apparatus, nervous system and metabolism make it unfit to survive. It has not attained requisite variety.

Exploring the nervous system model

There are many other aspects of this model specific to the various sub-systems. Here are a few.

Sub-system 1: actions

Are the actions at this level also sufficiently powerful to deal with actual or expected changes? What if demand for one of your products is growing at, say, 20 per cent a

---

9 This is sometimes called the 'bullwhip effect', after the way small initial errors and delays become violent swings further along the chain. This effect can readily be modelled and there is a widely used simulation based on it, called the Beer Game. The MIT Forum for Supply Chain Innovation has turned the game into a Web-based program. You can play it online, free, and either on your own or with colleagues, at http://beergame.mit.edu/gameapplet.asp.

year. If you can increase shipments by only 10 per cent, you are making room for your competitors to step in.

Conversely, if you respond by increasing supplies by 30 per cent, your overlarge inventory will soon start soaking up money and other resources. This possibly points to faults in intelligence-gathering (sub-system 4), internal coordination (level 2), decision or control (level 3). If your organization oscillates between too small and too great a response, it shows that it has slow reactions or faulty information supply.[10]

Sub-system 2: coordination

Dampening oscillations within the production system being studied is one of the tasks of the coordination sub-system. It should do this in a just-in-time (JIT) fashion, neither so early as to be forgotten or confuse nor so late as to be irrelevant. Actions at this level need to link with their counterparts in other nervous systems, such as in the operating divisions of an organization. These links help to minimize intraorganizational variation and competition. Coordination arrangements usually combine formality and informality. Some are spontaneously set up by the production units themselves, and are thus part of the unofficial organization.

Sub-system 3a: monitoring

Is internal monitoring relevant, timely and sufficiently frequent, as discussed above? Is data quality good enough? Is it comprehensible? Is the monitoring system's variety – its capacity to deal with information and change – greater than that of the sub-systems it is overseeing? If not, it risks being overwhelmed and thus passing on inadequate or distorted information to the management sub-systems.

This is a common fate, often stemming from oversimplified reporting mechanisms. Let us say, for example, that you measure only a division's output volumes and throughput rates but not customers' rejection rates as well. Such as reporting system would give you an unduly rosy picture of performance. It would not, for example, alert you to the fall in product quality caused by problems in the factory. Providing measurements of quality would add necessary variety to the reporting channel.

---

10 There are longer-term oscillations in organizations. One well known one is their tendency to fluctuate between centralizing power and resources and dispersing them. Another oscillation is in the influence and reporting position of the IT function. Is it regarded as a business-driven unit, reporting to the CEO, or is it seen as an overhead trimmer, reporting to the CFO? Neither extreme is sustainable.

Sub-system 3: control

Is it good enough to ensure that actions are effective and carried out efficiently? The control sub-system needs to be sensitive enough (possessing enough variety) to respond suitably to the information and instructions coming to it from levels 4 and 5. The instructions it gives to the operating units at level 1 must be expressed in terms acceptable to them and in a comprehensible manner.

If control is exercised too weakly, operating units will either become unruly or work at cross-purposes. If it is exercised too strongly, the level 1 units become demotivated and placed under stress. This is a common result of computer-based centralization.

Sub-system 4: intelligence

Is good enough information getting through to the organism? Is it up-to-date, relevant, reliable, timely and sufficiently frequent? Is it comprehensible to those who must make decisions based on it? What variety is lost in making it comprehensible (the problem of oversimplification again)? Is its internal distribution relevant, timely and sufficiently frequent? Is intelligence-gathering based on a good enough model of the organization and the environment, in order to know what to look out for?

Sub-system 5: policy

This is where the organism consciously plans actions in pursuit of its purpose. Is the planning done using adequate models, fresh information and a wide enough range of it? This is the place where thinking about the 'here and now' of sub-systems 1–3 is – or should be – combined with the 'there and then' information gathered at level 4. It is where most anticipatory – as opposed to reactive – actions are decided.[11]

The Viable System Model does not explain or deal with everything, but no model can. Matters such as people management, staff and personal development, managing change and product design are relegated to the background in this view. Despite this, it forms a helpful basis on which to examine the processes of an organization and the kind of computer system that helps with them. (For more on different ways of looking at organizations, see chapter 6.)

---

11 Not all actions are instigated at level 5. In animals, flinching is a semi-autonomous anticipatory action, based on earlier unpleasant learning. The higher brain has little control over it. In punitive organizations, an equivalent is perhaps the manipulation or covering up of data that conveys bad news. Rather than get into trouble from the next higher organizational level, local managers 'massage' the figures.

Table 4.2 The Viable System Model and software

Sub-system	Sample types of application software
1. **Actions**	Word processing, CAD, computer-aided manufacturing (CAM); process simulation and design
2. **Coordination**	Groupware, shared diaries, CRM, EDM
3. **Control**	Production control software, workflow automation
3a. **Monitoring**	ERP, BI, BAM
4. **Intelligence**	General and stockmarket news feeds
5. **Policy**	None

Where computer software comes in

Computer software, broadly speaking, is of four types. It:

- enables the computer system to run; this includes operating systems and software for database management, system management and networking
- helps human beings operate the system; examples include programming languages, graphical user interfaces (often provided with the operating system in personal computers) and World Wide Web browsers
- solves the problems that computers create, such as exchanging data among dissimilar systems, converting among different protocols for data communication and file creation and managing security
- does actual work, such as accounts packages and spreadsheets, and programs for graphics, word processing and production control.

There is, to the outsider, an astonishing amount of time, effort and money devoted to the first three categories. It astonishes – and dismays – many insiders, too. For the user organization the first three categories all count as overhead ('burden') or on-costs. Reducing their drain on corporate resources is an important objective for any organization that uses computers.

The last category, software that makes a positive contribution, is the focus of attention in this section. Table 4.2 shows is how the Viable System Model applies to types of application software.

As we hope is clear from our comments in chapter 3, BPM software straddles several categories and contains tools for sub-systems 1, 2, 3 and 3a. In other words, it helps organizations to coordinate, control and monitor their actions. As Jim Sinur of Gartner Group puts it: 'BPM offers what the ENS [enterprise nervous system] needs: rich coordination, a strong memory of processes and a process-focused view of business.'

BPM software also helps with developing new systems and stitching together old ones, but this is not an explicit part of the Viable System Model. Solving the problems that computers create is also part of the job of EAI software. We have therefore omitted it from table 4.2. It makes no overt contribution to any of the five sub-systems.

We discuss below four of the types of software listed in table 4.2 – ERP, customer relationship management (CRM), BI and BAM – and touch on EAI.

ERP

ERP originated in manufacturing companies, being first called materials requirements planning (MRP) and then manufacturing resource planning (MRP II). These days, ERP is found in many other sectors besides manufacturing and has grown to encompass several other corporate functions. Originally, it handled just the main phases of the manufacturing process – planning, getting, making and delivering. It now typically incorporates modules for plant maintenance, HR management, financial control and BI and modelling.

The ever-increasing use of ERP software since the 1980s is not just the result of astute marketing by its makers and by systems integrators. It also reflects the growing wish by industrial organizations for overall, integrated control of their operations. Users are constantly looking to improve in efficiency, speed and cost.

ERP helps with this by having an organization's various planning and control systems work to a single data model. Actions are then reported into and driven from a single database or a homogeneous set of databases.

The data-driven method of working that results from this architecture provides consistency of operation for the user organization. An ERP system can give internal developers a single set of standards to work to. Also, it passes upgrade responsibility for LOB application programs to the ERP software supplier. (See below for more on these.)

This all comes at some cost in upheaval, slow time to benefit and high licence and upgrade fees. Often, there are conversion problems when existing application programs are being replaced. Sometimes there are difficulties when trying to integrate existing programs with ERP packages.

There is an operational weakness, too. While ERP software is excellent at controlling events within its own ambit, it typically can do little outside that. Even within its own applications, the kind of process management it can do is limited in variability and in sophistication. (We discussed some of these aspects in chapter 2.)

In ERP software, if a particular process is not a standard part of a particular package, it must be 'hard coded' in. Because this is difficult, there is a temptation, often yielded to, for the systems designer to try to simplify reality. The full range of potential variations in destinations, points of origination, routings, timings and so on become whittled down to something more easily managed from a programming viewpoint. Naturally, this makes the operation of the process a bad fit with what the users must try to achieve. This leads to accusations, often justified, that the ERP system is rigid and out of date.

Specialized process management products, by contrast, can model and manage levels of complexity far in excess of those offered in most ERP products. They also require less programming effort to set up process models. Should the model embodied in the software be found not to be an exact or adaptable enough fit with reality, it is a relatively insignificant task to make any necessary modifications. Additionally, specialized process software can detect and manage a process regardless of where it originates, is routed to or finishes.

ERP software and its makers hold out to senior managers and their technical staff the promise of wrapping all their problems into one easily managed bundle. This kind of thinking goes back to the earliest days of corporate computing, with management information systems (MIS) then being the Holy Grail. The temptation to invest in such 'one big answer' systems is hard to resist for the uninformed and inexperienced. Exposure to reality is usually a good teacher. ERP systems are good at many things but are not, nor cannot ever be, a complete answer. No single piece or type of software can.

This has stirred up yet another integration versus 'best-of-breed' argument among software companies, as has the ERP vendors' targeting of LOB software. This term, a favourite with IBM, simply means software to help a company with 'the things it does for a living – e.g. banking, manufacturing, etc'.[12] These tend to be bespoke pieces of software, sometimes written by the organization's own programmers. Changing LOB packages is not for the faint-hearted; linking with them is often as much as can be done (also discussed in chapter 2).

To try to reduce the cost, complexity and intractability of their products, ERP suppliers have in recent years been offering them in sub-assemblies – several moderately sized answers, in effect. They have also begun looking to the World Wide Web as a means of delivering ERP functions to user organizations, possibly as a chargeable service. We discuss this further in Chapter 5.

EAI

One reason for these changes is the pressure that ERP software makers have been under from the suppliers of software or services for EAI. This is the effort to draw together, manage and control the organization's entire stock of application software and holds together many a corporation's systems.

EAI products are often referred to as 'middleware'. Middleware simplifies life for software developers working in complex systems by allowing them to write

---

12 From J. Hosler (ed.), *The Handbook of IBM Terminology*, Xephon, Spring 1995.

programs once only. Where there are diverse types of client and server system to be integrated, middleware contains within it the application programming interfaces (APIs) for each of them.

In a complex system, typically for distributed or inter-enterprise computing, a developer would normally have to program for each combination of client and server that exists in the system. Using middleware, the programmer has only to link the program he is working on to the middleware. It becomes a matter of 'write once, run anywhere'.

Middleware allows developers to treat the organization's software as a collection of independent yet loosely connected system components and services, rather than as a monolithic entity. It is simultaneously glue and elastic, allowing previously separate programs to be linked and previously centralized programs to be dispersed around the organization or outside it. BPM software acts as middleware in many respects.

Much as one would expect, ERP vendors dispute the need for EAI (while secretly buying companies providing software for it). As usual in computing, the forces of centrism are fighting the forces of specialization. This is an endless and endlessly repeated struggle, which for computer users often seems only to produce more confusing terminology.

CRM

This is the use of computers to help organizations deal with potential and current buyers of their products or services. CRM systems are typically built around a database of customer and trading details, updated from the organization's systems and elsewhere. Organizations use these systems to help them identify and target customers, produce sales leads, improve account and sales management and run marketing campaigns.

Despite the name, CRM systems do not directly manage the relationship with the customer. Rather, they manage the *data* about the customer, usually these days tying it in with data about products, production, markets and trends. Their original purpose was as an aid to salespeople, helping them work in a better-informed and coordinated way. CRM began as contact management programs, keeping records of sales and previous customer contacts and prompting repeat calls through a reminder calendar. These packages evolved into salesforce automation systems, allowing salespeople to share information. Full-blown CRM goes even further, linking to call-centres and technical support teams. Like most modern application programs, it can work over mobile networks.

CRM systems lie behind the sales activities of most medium or large commercial organizations. They help not only with human selling, either face-to-face or through a call-centre, but also that done through electronic channels such as the World Wide Web.

This allows a degree of personalization of online communications with customers. A Web site, for example, can be made to look different for each customer each time he visits. The site can present information relevant only to the customer and, perhaps, will make special price offers in areas of his previous or declared interest.

Data-driven personalization is part of a trend to one-to-one marketing. The aim is to treat the customer as an individual even though there may be no or little human contact involved. Online booksellers and grocery chains use this ability extensively. Larger CRM systems rely on some form of data mining to help tune their actions and responses.

Modern CRM software often includes some workflow automation. Using a combination of triggers and rules, it can, for example, assign prospective leads to salespeople in the right geographical area and with the relevant abilities. These process tools are often rudimentary and work only with the CRM system itself. Newer CRM systems are gradually adopting the same standards as BPM products. This accompanies an increasing trend to break down CRM software into modules, as with ERP systems, and to make those available as a service.

BI and BAM

If you will sit down with me in my office for twenty minutes, I will show you what the condition of business is at any given time in any locality of the United States.

So boasted William Orton, president of the Western Union telegraph company in 1870.[13] Modern managers do not have twenty minutes to spare – they want quick answers – but they want the kind of information Orton was promising.

In computing, BI is the supply of online numerical information extracted from an organization's data stores. As well as historical analyses, it includes projections into the future based on 'mining' that data and occasionally adds external information. The information supply is directed mostly at managers and directors (executives), who use it when making tactical and strategic decisions. It thus equates to sub-systems 3a and 4 in the Viable System Model.

An earlier name for this sort of service is enterprise, or executive, information system (EIS). Increasingly, an organization's customers and trading partners can also tap into the information flow.

Although helpful, these systems do not provide users with current knowledge of what is going on in their areas of concern. In the jargon, they are data-driven not event-driven. The information they present is not fresh enough to deal with operational or short-cycle tactical changes. This is seen as a drawback in an age when the

---

13 Quoted in Tom Standage's book, *The Victorian Internet*.

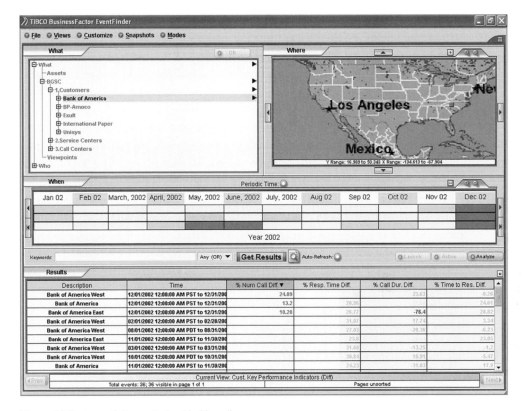

Figure 4.3 Typical activity monitoring 'dashboard'

emphasis is on organizational nimbleness ('agility') and speed of response. These days the drive is to what consultants like to call the real-time or zero-latency enterprise.[14] Another favoured term is the event-driven enterprise.[15]

14 Latency, or latent period, is the delay between a stimulus and a response. It is a measure of how much a reaction lags behind an input and reflects the inertia of the system or organism being controlled. An information or control system might respond quickly but that does not mean the entity it is working into will be as sprightly or alert. The latency of a computer system is many times shorter than that of the organizational unit it services, for example. It is the overall latency that matters. The 'zero-latency enterprise' is thus a fine ambition, but as unrealistic as perpetual motion.

The 'real-time enterprise' is equally impossible to achieve. Genuine real-time systems control nuclear reactions, fighter aeroplanes and similarly volatile arrangements. They react in thousandths of a second. The kind of organization-wide management system discussed here changes an organization's behaviour much more slowly. Doing so will take hours in exceptional cases, days sometimes and weeks usually. That is probably much faster than before, but it is still not 'real' time.

15 To our mind, this expression does not even suggest a worthwhile objective. It carries unfortunate connotations of being pushed hither and thither by circumstance, of being directionless and of acting reactively rather than with foresight. Create event-driven systems or sub-systems by all means, but the overall organization must be goal-seeking.

The reporting tools in BPM give this up-to-the-minute detail, in a service increasingly being referred to as 'business activity monitoring' or BAM.[16] These tools connect to the BPM system and to the application programs it manages or keeps a watch on. Not all programs need be under direct BPM control. The BAM element can still watch what these programs are doing and trigger a process that reacts to any exceptions arising.

If a watched process is wholly or partly under human control, then the BAM tools can tell those people the state of affairs and possibly suggest suitable action. Typically this would be by email or as a workflow task.

The links that supply the necessary data are plumbed in when setting up BPM. This can be part of an EAI project elsewhere in the organization.

Figure 4.3 is an example of the kind of screen that a user would see. It is often referred to as a management or enterprise dashboard, from its resemblance to the instrumentation on an automobile. This example is of TIBCO's BusinessFactor software. It shows several displays that between them give 'what', 'where' and 'when' views of performance data. At the bottom is a highlighted spreadsheet, with negative values in red (bold). If a user sees an indicator moving too far or too fast, or receives an alert from the system, he can launch corrective action. The user can also ask for more or different data. In some products, he can simulate possible scenarios, a visual equivalent to the 'what if' calculations in spreadsheet programs.

BAM sensors connect not only to data stores and the BPM system but also to network and computer management systems. They can be set to detect, for instance, network failures, database access loads and Web site activity.

The amount and type of information presented will vary according to the user's needs. Somebody monitoring a highly integrated and automated business process will want to know immediately and in detail if any part of it breaks down or starts running outside set limits. A board-level director, by contrast, will perhaps want only a daily digest of these happenings, being more concerned with company-wide results. Details of any breaches of service-level agreements or business governance rules are typical requirements at this level.

It is an important part of the overall BPM task to define what these dashboards show to whom, how the data is filtered and analysed and what the response mechanisms should be. The concept of requisite variety applies with great force at this point. Also, as with any design activity involving users, this stage is best carried out progressively and with users' intimate involvement.

In chapter 5, we look at how all this affects the world outside the organization and its dealings with it.

---

16   It is also sometimes called business performance management (another BPM) but we will stick with BAM.

Case Study 3 International Truck and Engine Corporation

International Truck and Engine Corporation ('International') is the main operating division of its parent, Navistar Corporation, and provides around 95 per cent of its sales income. Like any manufacturing organization, it needs to document and, where possible, expedite changes to its products. Its old system for doing this used paper forms, which moved around inefficiently and in an uncontrolled way. This was a similar problem to that in Canton Zug (chapter 2) but much more complicated and involving greater volumes of data.

International's chosen solution was to apply a workflow product that it was already using in other parts of the organization. This aided learning on the part of users and systems staff alike. Also, given the industry they are in, many employees probably already possessed a process-oriented viewpoint. Nonetheless, they still needed to learn how to translate that into a specific system design and learn how it would affect their daily work.

The company involved potential users closely in the design and prototyping of the new system. It also ensured that senior managers gave a clear lead on the project and support for it. An important outcome of its design methods was having a clearly recognized owner for each business process.

There are excellent productivity and financial results arising from International's adoption of this process management system. Among the human results have been greater feelings of accountability among users, a release from constant 'fire-fighting' and improved relations among the various departments involved.

Product Change Management System (PCMS)

Industry/ Sector	Automobile manufacture	Location(s)	USA
Annual turnover/ income	Not available	Number of employees	Over 14,000
Type of system	EDM/workflow	Supplier and product	Action Technologies ActionWorks
Number of users	Nearly 1,000, in many plants in 3 countries	Time to complete	Various pieces of 3–9 months each
Business objectives	Automating the management of product changes		

Quantitative results	• ROI of 362 per cent
	• Paid for itself in 3 months
	• Productivity up 30 per cent
	• Cycle time down 75 per cent
	• Rework down 33 per cent
	• Application development costs down 30 per cent–50 per cent
	• Training costs halved
Qualitative results	• Managers feel greater accountability
	• Process data is captured automatically
	• Immediate implementation of process changes

Business background

International Truck and Engine Corporation is the $7.3 billion dollar operating division of Navistar International Corporation. It has the USA's largest combined market share in medium-to-heavy-duty trucks. World-wide shipments in fiscal 2003 totalled 82,200 trucks and buses. The company shipped 396,000 engine units in 2003.

The truck manufacturing industry world-wide is going through the same pattern of stresses and change as did car manufacturing in the 1970s and 1980s. This involves globalization, consolidation, cost-cutting, supply chain streamlining, faster product innovation and a new responsiveness to customers.

International is replacing four-fifths of its models, using shared modular components where possible. The company is also moving to 'exactly in time' manufacturing. Major elements of this are shorter product development cycles, lower production costs and newly designed vehicles built to 'car-like' quality. It aims to be profitable on low volumes.

BPM improvement is a important contributor to this strategy. Part of it has been a successful Six Sigma quality programme. International also has a new process for product development, to speed the introduction of new truck models, reduce rework and sharply cut costs.

Within this, the company developed a new product change management system (PCMS). The old process was widely recognized as fundamental to the company but also a source of problems. It was paper-based and lacked standardization. Also, it was confusing because it was so complex. There were too many 'handoffs' from one group to another. Many different departments created isolated ways of dealing with change orders, which needed troubleshooting and negotiation. Work-arounds abounded.

Employees had no clear understanding of the entire process. Critical process information and performance measures were seldom recorded. Where they were, they remained in disparate data sources and spreadsheets on any of several hundred computers. Team members never knew at any moment who had undertaken to deliver a specific part of a product change, who was working on what or when it was due to be completed.

International selected Action Technologies' workflow automation product to help it create the new PCMS. The company's Process Development Department at the Truck Development and Technology Center (TDTC) was already using Action's product for several automated, Web-based administrative processes. Bill Bailey, its Process Development Department Manager, first saw the potential benefits of workflow automation in 1999 and has championed its use ever since.

System description

The system crosses many functional areas. A change request from the Product Center first goes for business approval. It then moves to programme staffing, new component procurement, engineering design and delivery, quality assurance, production planning and then to final release. The result is a completed order for a new feature. This process involves thousands of people and creates more than 10,000 change orders a year, involving more than 100,000 product elements.

Underlying the process is the Change Request. This is where all the ideas for product change and innovation are created and debated among various department personnel.

Next is the Change Proposal, which moves the process from conception to planning. All the ideas from the Change Request are electronically routed for review, then approval or rejection.

After this comes Change Development. This includes authorizing the work and the full design, development and engineering release of the change order. The order mobilizes the resources needed for implementation. Work in this phase involves intense negotiation.

Change Implementation is where the change is put into operation. It varies in time and cost, depending on many factors. These include the nature of the components and whether this is an improvement, redesign or a new design.

Action Technologies' ActionWorks software lends itself to this sort of negotiated process. It coordinates interactions between an individual or group making a request (the customer) and the recipient of that request (the performer) in four phases:

- preparation – the customer proposes work to be done by the performer and issues a request
- negotiation – the customer and performer negotiate until they agree ('commitment') on the work to be done

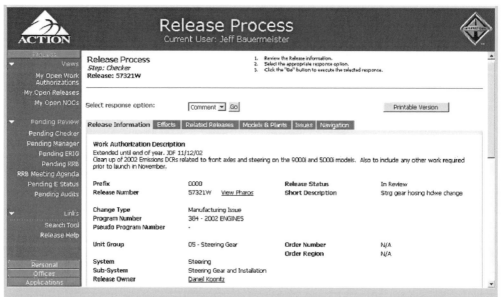

Figure C3.1 PCMS work authorization screen

- performance – the performer fulfils the request and reports completion
- acceptance – the customer evaluates the work and either declares satisfaction or points out what remains to be done to fulfil the request.

The PCMS uses ActionWorks Metro and Process Manager software, running on a Compaq server. Another Compaq server hosts Microsoft SQL Server 2000 and Internet Information Server. These systems link to International's Oracle Database and IBM DB2 databases. Workstations access PCMS through Microsoft Internet Explorer (figure C3.1).

International has standardized on Action Technologies' ActionWorks BPM software. Today, the company has twenty business processes running on the system, with more in development.

Implementation experience

Where and when did the project or system originate?	From within International's Truck TDTC. Workflow automation was first seen in 1999
How long did implementation take?	Individual processes were developed in 3–9 months with teams of 3–12 people
Who did the implementing (own staff, contractors, consultants, etc.)?	Done using internal resources to retain knowledge and provide day-to-day support and future enhancements

How much bespoke development was there?	All of it was bespoke, using the Actionworks tools, HTML pages, ASP server code, SQL, DB2 and Oracle databases
Were there any special infrastructure needs?	None; the system uses the existing Web-based infrastructure, Microsoft Internet Explorer and SQL server
What were the most significant implementation issues and how were they dealt with?	Finding process owners and defining the processes to the correct level to provide the right information at the right time to the right person; this meant learning new skills, as this was the first true business process to be automated within the area and differs markedly from creating a database application
Who is responsible for the system overall?	There are dual responsibilities: • The Process Development Department and IT Department jointly provide system support • The Process Owner is responsible for managing and approving changes to the process
How was and is training handled?	Great care was taken to give all processes a common look-and-feel; this allows users to exploit their previous Web and process knowledge; most training occurs in large conference rooms, with specialized training as required for specific groups on their piece of the process
What was and is done to encourage use?	Everyone must use the automated process
What lessons were learned?	• Each process must have a strong process owner who controls changes to the process • Process ownership is hard work but the benefits are large • Automating the business process provides a leap in productivity rather than an incremental improvement

International split the work of improving the process into separate projects, each lasting from six to nine months. Creating the Change Development part of the process was the most complex of these. Neal Cunningham, Process Development Manager, assembled a team of eight people, consisting mostly of business users. In a series of meetings, he and the team roughed out a prototype in a few weeks, using the ActionWorks Business Interaction Model.

The team asked such questions as:

- Exactly what work is required?
- Who is asking for the work to be performed (who is in the role of the customer)?
- Who has to do the work (who is in the role of performer)?
- When is the work due?

There were remarkably few hurdles to overcome in any area of management. Mainly this was because the management team at International's TDTC fully supported the initiative. Also, everyone using the old process saw the need to improve it.

The Process Development Department did not have to 'sell' BPM to the management team. Their earlier experience of using the ActionWorks methods and software had shown managers its abilities.

Members of the Process Development group and the IT departments worked closely together on the project.

Users were told about the plans for and design of the system well in advance. Near the end of implementation, they were given informal briefings on the tool and how it works. Formal training followed this.

Benefits and user reaction

What has been the reaction of managers and staff?	• Easy to use • Makes information easily available
What has been the reaction of customers or trading partners?	Customers are primarily internal users located in many plants in three countries; positive feedback is the norm, with requests for additional workflow automation
What has been the overall cost of the system?	3–5 process development professionals and 2 IT professionals, over several years; the initial cost and continuing maintenance cost of the ActionWorks software was small in comparison to the cost of the professionals
What have been the main process benefits?	• Consistent, clearly defined Web-based processes with consistent similar interfaces provide the right information at the right time to the right users • Better communication in general • Virtual offices have allowed work to be managed regardless of where people reside
What have been the main effects on operating style and methods?	• People can focus on their job and let the process be the administrator • Everyone follows the same process without shortcuts or work-arounds

PCMS allows all participants in the change process to coordinate their efforts. Team members can now see what is happening at every step. The system has eliminated almost all complexity and associated delays, costs and errors.

Four hundred engineers access PCMS every day. Here are some of their comments on it:

1. 'I have 28 years' experience in the company. As a programme manager, I'm always being asked what the status of a job is. Now, I have a complete record of

everything via one screen: status, feedback, emails about completion. I can track programme completion versus promises with minimal time and no effort. I track work in progress and contact engineers to find out why there are delays.'

2. 'The system has changed our weekly meetings with the other business functions. As engineering representatives, we used to spend most of our time on questions about jobs. Now, the topic on the table is the business – purchasing, service, and cost analysis. I've generally been the person in the middle of a lot of chaos. I used to take a copy of the job list with me into the meeting. Now, I'm forewarned about any problem areas. I'm better prepared and far more confident.'

3. 'I now know the job inventory across three countries, which I never knew before. When I was visiting our Mexico operations, it was very hard to keep track of what was going on back home and vice versa.'

4. 'I have far better relationships with the engineering community. There is more dialogue and time for trying out new ideas instead of getting stuck on dates and status.'

5. 'There's no more finger pointing. It takes a lot of noise out of the process.'

Managers uniformly report that PCMS improves their personal productivity, interactions with other parties and business contributions. In addition, they point to a new degree of accountability for themselves and their colleagues for meeting commitments. They see this as an important step towards International's transformation.

Bill Bailey, head of the Process Development Department for TDTC, says the system has helped the Engineering Release Integrity Group move from being 'firefighters'. They now play a more anticipatory role in problem solving and planning.

The results of introducing the new systems are evident in three directions – lower costs, shorter time-to-market and increased quality:

- 362 per cent return on investment; three-month payback. Despite these outstanding results, Tom Wilson the Controller at TDTC cautions that ROIs are company-specific. 'When looking at case studies of possible tools, systems or methods to buy, we are much more interested in where we can improve efficiency in International's tools, systems and processes rather than another company's ROI. This is because of the company-specific assumptions necessary to calculate them are so different from company to company.'

- 30 per cent increase in management personnel efficiency. These are typically the most experienced and highest paid workers in the process.

- 50–70 per cent cut in development costs. ActionWorks allows ordinary users to carry out most of the process design work. This allows the company's scarce computer staff to concentrate on more technical work. A side benefit has been creating the company's own consistent and easy-to-use process recording system. Users have adopted this.

- 60 per cent of variation change requests are now rejected early. All product change requests are now evaluated early in the change process. Team members can easily winnow out changes that are too expensive or which offer little value to the customer and these are rejected.
- Halving of training costs. PCMS' Web-based tools all use the same general user interface and documentation. Each new application therefore builds upon the users' previous training and knowledge. Training averages one–four hours, depending upon the application. 'Where else can you have participants go through only a couple hours of training to learn how to utilize a new system for managing a core business process, and be fully functional? People just naturally took to it', says Jeff Bauermeister, Process Development Manager.
- Eliminating unauthorized work. All work is now negotiated, approved and, when completed, accepted by the product centres. The company estimates that a tenth of previous work would not have been authorized or would have been combined into other product change efforts.
- Reduced time-to-market. Cycle time improvements of 60–75 per cent were achieved in two important steps of the process. This represents a significant competitive advantage to International.
- Rework cut by a quarter. The process design team moved an automated error checking routine forward in the Engineering Release process. This and improving the types of errors caught before sending the work has resulted in 33 per cent more errors being caught than before.
- Process data capture. PCMS records all process data. This can be analysed and used as a baseline for future process improvements.
- Immediate process change implementation. Any changes to the process are immediate, which has a direct effect on quality, cost and cycle time.

Although all its main competitors are also racing to develop new products, International is now a leader. Without PCMS, this radical and dramatic improvement would not have been possible. Phil Christman, Vice-President of Product Development at International, is a frequent user of PCMS. He says: 'We are dependent upon these processes today and we look forward to future improvements to increase our productivity and speed to market.'

The future

International continues to improve its product development processes. The company is embarking on a major project to improve its overall Development Chain Management (DCM) and product life cycle management (PLM) processes. DCM is a 'best-in-class' approach to integrating the core business processes behind product development, such as portfolio, pipeline, financial and resource management. PLM is the management of the information and decisions made about a product or service offering, from initial concept to end-of-life. The Product Change Management System is an important component of these processes and will likely see additional development, expansion and integration over the next few years.

5 Looking outside the organization

In chapter 4 we looked at software that handles the data flows that arise within the user organization. Managing the flows *between* organizations falls to another category – supply chain management (SCM) software. This deals with moving goods, data and money (as data) between trading partners and out to the customer.

Many of the costs of any supplier or manufacturer are tied up in the supply chain. This makes the detailed and responsive management of buying and inventory of major importance. Being able to forecast demand and supply is also critical. Tom McGuffog, a former director of electronic business with Nestlé UK, sums up the problem with this pun: 'Uncertainty is the mother of inventory'.[1]

It is no surprise, then, that excess inventory is one of the primary targets of the users of SCM software. Inventory hides deficiencies: exposing and correcting those deficiencies is a goal for any trading company. The ERP software makers have increasingly looked to extend their products into the inter-company realm to help with this. As a result, the distinction between the technologies of ERP and SCM, and their consequent ability to deal with supply chain matters, is becoming blurred.

Managing the supply chain

Although the metaphor of a supply chain is of late twentieth-century coinage, the thinking behind it as old as mass production. Unless occupying a special niche, manufacturers always want speedy, reliable and inexpensive movement of raw materials to the factory. They also want a similarly efficient movement of finished goods to the customer, either direct or via an intermediary. (The latter journey is often, and logically, called the demand chain. For simplicity, we use 'supply chain' to mean upstream as well as downstream activities.)

---

1 Quoted in a presentation McGuffog made to the UK Council for Electronic Business in February 2003.

Using computers to help with either chain was a natural step and dates from the early 1960s.[2] Electronic data interchange (EDI) became the staple method for this, the first set of standards for it arising in 1978. By today's expectations, EDI is slow, rigid and unresponsive but, in its favour, it is widespread, stable and well understood.

Since the early 1990s, much attention has been paid to using the Internet to speed these external data flows. The rise of the World Wide Web as a trading medium prompted the creation of new methods for exchanging and using information. Some people heralded this as a revolution to rival the rise of railways or the telephone. As it turned out, it often resulted in unrealistic and, mostly, unrealized expectations. Only a few of the so-called dot.com organizations, such as Amazon, eBay and Yahoo!, survived for long. (Google came later, building on the efforts of pioneers such as Alta Vista.)

Nonetheless, electronic commerce (or e-commerce) has made a significant difference to existing organizations, but often not directly. Tesco, the British supermarket chain, provides an example. Tesco.com is probably the most world's most successful online grocery site, making online sales of £577 million in the year to February 2004. This is a handsome amount of business but is a mere fraction – just over 2 per cent – of the company's overall UK sales in the same period. These totalled £26.9 *billion*.

You might imagine from this that the Internet matters little to the company. The reverse is the case, as it is with many manufacturing, retail and service suppliers. Tesco relies on the Internet to manage its entire supply chain, communicate with its stores, logistics operations and suppliers, run its tills, link warehouse systems and so on.[3] A communications breakdown would paralyse the company even more quickly than any transport problems might. The same applies to other exemplars of Web-based 'bricks and mortar' businesses, such as Dell and Cisco.

For these organizations and others like them, the Internet is not just a communications highway. It is also the way in which they extend their computing activities beyond their boundaries. It is, these days, the essential foundation for their trading activities.

Trading activities

Before looking at electronic trading in more detail, let us examine the ordinary kind. What traders do can be split into five main activities:[4]

---

2 In the first known direct transference of trade data between computers, E. I. Dupont de Nemours began sending online cargo data to Chemical Leahman Tank Lines. Using electronic networks to transfer trade data between *people* goes back to the early days of Telex, in Victorian times.

3 Some commentators call this all-embracing approach 'value chain management'.

4 Manufacturers have also design and development chains, through which they bring to market new and revised goods and services. (See, for example, the International Truck and Engine case study in chapter 4.) Managing these involves similar principles as for demand and supply chains.

- the prelude to a transaction
- agreeing a contract
- fulfilling the contract
- getting paid (settlement)
- following up the transaction.

Buyers undertake a matching set of activities but, obviously, receive rather than deliver goods or services and make instead of receive payments.

The *prelude* to a transaction is all those activities that are an immediate preliminary to it, such as direct marketing, selling, dealing with enquiries and issuing quotations. It also includes effecting an introduction, such as by a recruitment or an estate agent. For buyers the build-up consists of specifying, 'sourcing' (that is, getting) and pricing. At this stage, there is no commitment on either side to trade.

That comes after negotiating a *contract* to deliver or buy goods or services. This is normally a distinct process in business. It is also separate in large-value consumer transactions, such a buying a car, a house or a pension. In most other forms of consumer trading, the contract is implied and is typically covered by legislation. The 'consideration' that is an essential part of any contract to trade is usually in the form of currency, but could involve barter or tokens.

Fulfilment is the delivery of the goods or services specified, in the agreed manner and timescale. In traditional commerce, some form of physical movement of people or objects is implied at this stage, even if it is just of shipping documents. In electronic trading, fulfilment of digital goods such as software or publications can be performed online. If what is traded is physical or needs direct human contact, fulfilment can at least be started and supervised on-line.

Settlement – the making or collecting of payment for goods or services – is, of course, the main point of commerce. It is not necessarily synchronized with the other activities described here; many suppliers provide credit terms of 30 days or more, while others demand prior payment.

For sellers, the *aftermath* of a transaction consists of activities such as post-sales service, warranty dealings and providing consumable supplies. For buyers, it involves tasks such as bedding in a new system, adapting it, maintaining it and, eventually, replacing and disposing of it.

Figures 5.1 and 5.2 show the sequence for buying and for selling, with some typical tasks under each of the five activities.

This sequence – prelude, contract, fulfilment, settlement and aftermath – applies equally in electronic trading. The difference is that these activities take place remotely, over an electronic channel. The chosen channel is not necessarily the Internet or even digital. It could equally be the telephone, Telex, videotex (for example, the French Minitel system), facsimile transmission or cable television. Few of these are presently digital from end to end.

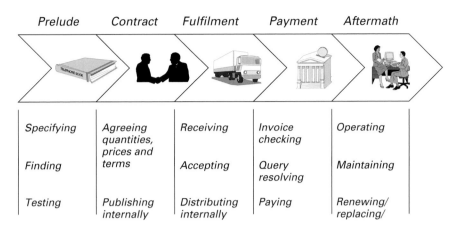

Figure 5.1 Main buying activities

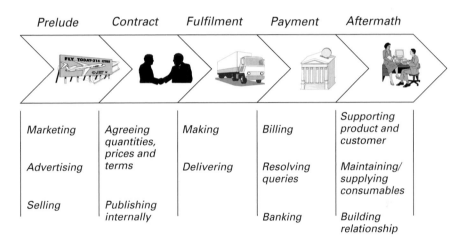

Figure 5.2 Main selling activities

Looking to the (near) future, the channel is as likely to be by wireless communication as by landline. The growing interest in wireless point of sale (PoS) systems and in radio frequency identification (RFID) devices is testimony to that.[5]

Figure 5.3 shows an idealized version of electronic trading, with each activity in the buying organization linked to its pair in the selling organization.

Common to all these actions is their aptness for attention in any programme of BPM. To varying extents, all involve computer-managed (or computer-manageable)

---

5 RFIDs are tiny radio transmitters designed to be embedded in products or their packaging. Computers can keep continuous track of their movement. Increasing numbers of organizations are using these devices in internal and external stock control and other mobile processes.

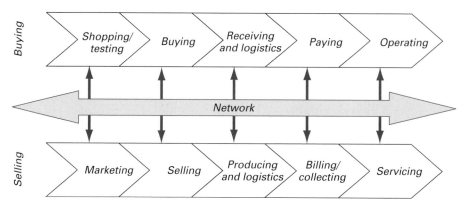

Figure 5.3 Buying and selling linked electronically

processes. As we saw in chapter 4, even where a BPM system cannot manage processes direct, it can at least do so at a remove. Keeping a watch on external processes is also possible wherever they are or whatever software drives them.

Who's trading with whom?

Most press coverage of electronic trading is about business-to-consumer (B2C) and business-to-business (B2B) trading. These are not the only two kinds of trading relationship, whether in traditional or electronic commerce. Public sector organizations are also highly active traders, on both the state's account and to meet their own needs.

The public sector also deals with the individual citizen, a consumer of government services. Unlike a consumer's relationship with the private sector, though, the trade flows both ways. In return for government services – a public 'good' – we pay taxes, duties and licence fees. Some people also receive grants or benefits payments.

Internal trading is another activity that receives little press attention. It may be hard to believe but there are some organizations that are their own biggest supplier and own biggest customer. Such large volumes of internal trade are characteristic of public and quasi-public organizations. In Britain, the British Broadcasting Corporation (BBC) and the National Health Service (NHS) are prime examples. In the private sector, internal trading is also prevalent in large, divisionalized companies.

A further important trading relationship is that between individuals. Consumer-to-consumer trading has long been recognized but also often been stigmatized as part of the 'black economy'. Increasing amounts of this kind of trade are now handled over the Internet, where they are usually seen as legitimate. eBay provides the main exemplar of this.

The result of these various relationships is a more complex pattern of trading than is popularly considered. Transactions take place among:

- commercial organizations – who could simultaneously be customers for, collaborators with and competitors to each other
- parts of the same organization – business-to-self
- public sector organizations or 'administrations'
- public and private sector organizations
- either of these and individual consumers
- individual consumers.

To complicate matters further, any of these trades can be made direct or via intermediaries. Whether labelled wholesalers, resellers, forwarders, retailers, agents or something else, these organizations and individuals interpose themselves between supplier and user. They do so usually to the benefit of one or the other (and to themselves, of course).

Contrary to mythology, intermediaries did not going away during the rise of the World Wide Web. Amazon.com is just a bookshop, not a large-scale publisher; eBay is an auction house, not a manufacturer of goods; Tesco is a retail grocer, not a farmer or grower. All are middlemen.

The image of a chain – linear, single-threaded and with links welded together – belies the breadth, complexity and dynamism of such a trading networks.

Electronic trading mapped

This set of linkages we describe is actually simple to picture: (figure 5.4).

The central four-blob map multiples and ramifies endlessly, creating a sort of picture of the virtual organization. Yet, as in the trading sequence discussed earlier, what the diagram shows is no different in essence to the relationships that obtain in traditional commerce. What is new in electronic trading are the reduced opportunity and operating costs, dramatically higher speed and near-ubiquity of the channels between these trading entities:

Every company has a diagram of the universe in which they're the centre. That's never true. We're all a node in a mesh. Intel's customers are customers of each other, our suppliers are customers of each other ... We all circle back to buy things from each other. It's an extraordinarily interdependent ecosystem we're a part of.

So said Douglas Busch, CIO of Intel Corp, in 2002 (quoted in D. Aponovich, 'Era of the E-Business Ecosystem', *Datamation* magazine). He put into words what we have tried to describe pictorially.

Figure 5.4 also depicts one of the targets of BPM (except, perhaps, in the consumer-to-consumer link). BPM allows us treat the organization as a rapidly responding network of resources and actors. This network extends beyond the company's legal boundaries to suppliers, trading partners and customers. These

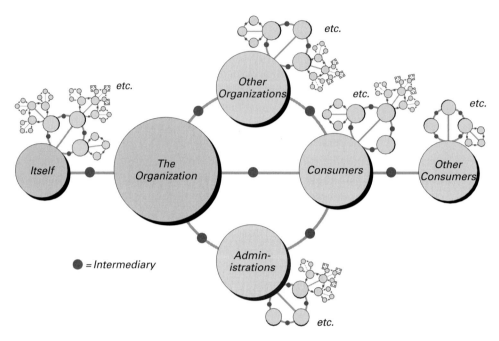

Figure 5.4 The electronic trading landscape

participants can be anywhere in the world at any particular moment, using both fixed and mobile communications systems to connect to one another.

One of the most widely used analysis tools in supply chain management is the Supply Chain Council's SCOR (Supply Chain Operations Reference) model. This divides the manufacturing supply chain into four main activities – planning, getting, making and delivering. Each of these can be analysed at multiple lower levels. Typical targets for analysis are a process's definition, its inputs and outputs, measures of performance and the systems and tools for enacting that process. All these are, of course, what process management software deals with so well.

AMR Research has coined the expression, 'demand-driven supply network' (DDSN), to describe a supply chain where action is energized and synchronized by what happens downstream of the manufacturer or supplier. The processes, infrastructure and information flows serve the downstream source of demand rather than an upstream source of constraints. It is the interorganizational equivalent of the *kanban* system for JIT production that Toyota introduced as part of its lean manufacturing philosophy.[6]

---

6 *Kanban* is Japanese for 'visible signal'. Typically a card attached to a container or bin, a *kanban* tells a supplier that the next batch of materials or parts is needed. The supplier can inspect the card directly or over a computer link.

Whether demand- or supply-driven, a supply chain relies on communication, coordination and control for its efficient operation. These 'three Cs' are what BPM is perfectly designed for. The sensitive and rapid activity monitoring implicit within BPM, for instance, permits better linkages between trading partners. It permits them to see undistorted and accurate demand information in time react to it, and gives them the tools to react appropriately. Perfect order fulfilment – with orders being delivered completely, accurately and on time – becomes practicable and affordable.

What happens at the systems level?

'The purpose of a business is to create and keep a customer.' So said Theodore Levitt in his book, *The Marketing Imagination*, required reading in many business schools. Doing so rests upon three main activities:

- Making and maintaining a point of sale that first-time as well as repeat customers will want to buy from. That point of sale can, for example, be a salesperson, a shop or, indeed, a Web site.
- Arranging matters so that the business always keeps the promises made at the point of sale. This means adequate product quality and prompt and accurate fulfilment, backed up with a good customer support operation.
- Creating and maintaining the underlying systems that make the two previous activities sustainable. Back-office systems must be efficient enough for the organization to be able to continue operating. (That does not necessarily have to be done profitably. Several high-profile Web-based businesses, for instance, ran or have run at a loss for years, but it has usually been at a planned-for loss.)

These tasks – attract, deliver and optimize – were not new with the Web. As with the models discussed earlier, it is what all trading organizations do anyway, with or without the use of electronic systems. A move into electronic trading simply changes the means for doing so.

The computer systems used at any of these three levels should integrate with each other, if only at the data transfer level. Figure 5.5 is a simplified diagram of a three-part trading relationship.

These three organizations form a microcosm of the kind of network of inter-trading organizations that make up the world of commerce. We show each organization as containing three kinds of system within it – SCM, ERP and LOB. (Note that these systems do not directly match the 'attract, deliver and optimize' model. Software is typically sold in packages that suit the production and marketing convenience of its makers rather than the needs of their customers.)

Linking these three organizations' systems is easier and cheaper if the protocols and data formats for interworking between them are standardized. So far, there are only a few standards that that have been universally accepted but there is progress

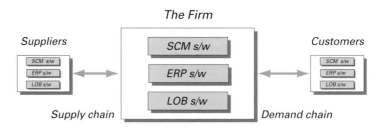

Figure 5.5 A trading trio

towards creating more. Most of this standards work is based on schemes and methods that use XML.[7] An example is ebXML (Electronic Business using eXtensible Markup Language). This is the outcome of a collaboration between CEFACT, a United Nations agency, and the standards group, OASIS. ebXML is a general framework for electronic trading that will eventually supplant EDI.

There are specialized uses of XML, too:

- In insurance, for example, the international ACORD (Association for Cooperative Operations Research and Development) group has produced XML-based communication standards for life and annuity, property and casualty, and reinsurance business
- Information in management and financial accounts can be readily exchanged, internally and externally, by using XBRL (eXtensible Business Reporting Language)
- Most electronic forms products (see chapter 3) are now using XML to link forms to process management servers.

Governments are moving to XML. In Britain, the UK e-Government Interoperability Framework announced in 2002 is based on it. Other national governments adopting XML include Germany, Hong Kong and New Zealand. Several American government departments have adopted XML-based standards for internal and external communication.

Process-enabling the supply chain

The process interchange standards created by the Workflow Management Coalition (WfMC) originated before XML was devised but now use it extensively.[8] At the

---

7 XML gives software makers a straightforward, adaptable way to create common data formats. These, and the data they contain, can be shared over the World Wide Web and other networks. XML is related to HTML, the basis of most Web pages. HTML is designed for displaying pages for human use and is not up to the task of handling large amounts of data, which is why XML was developed.

8 WfMC is a non-profit, international organization of workflow software makers, users, analysts and university and research groups. All the world's major suppliers of process management software are members.

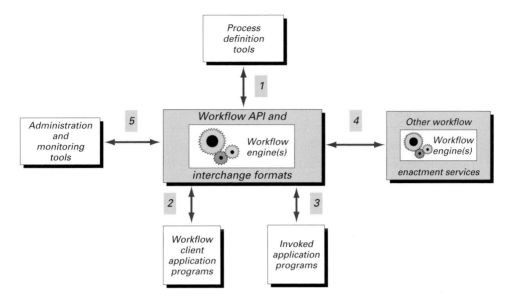

Figure 5.6 WfMC reference model

centre is its reference model. This offers standards for the interfaces between the elements of a process management system.

As figure 5.6 shows, it caters for multiple servers, which can be anywhere. Those servers can be of the same or different makes and can be in the same or different organizations. (Note that where the WfMC says 'workflow', you can safely read 'BPM'.) These interface standards are in daily use in thousands of organizations and work between the products of dozens of suppliers.

The five interfaces do varied jobs:

- Interface 1 links software tools such as process modellers, simulators and designer suites to the software that will run the process or processes (the 'engine').
- Interface 2 connects the engine to software running on access devices such as personal computers. It transfers data and commands between the users' machines and the server. The display of workflow queues is a typical example, as is the actioning of users' instructions.
- Interface 3 allows the engine to call up other application software, anywhere, for related tasks. These could include looking up an accounts database or triggering the despatch of email messages. (Although shown separately, in practice this is usually combined with interface 2.)
- Interface 4 is the main interface for cross-organizational working. It lets one process engine talk to another, to swap tasks perhaps or synchronize actions across a supply chain.
- Interface 5 provides a standardized way for system management programs to oversee the activities on the server. BAM software would use this link.

Using software that conforms to these standards helps an organization extend its process control and monitoring outside its boundaries, so long as its trading partners have done likewise. If it and they have not, making those links will require a great deal of bespoke computer programming ('custom code'). This, as we have mentioned, is slow, expensive and hard to replicate. Avoiding the need for custom coding is, of course, one of the main reasons for adopting BPM.

BPM not only provides a common system for moving information up and down a supply chain. It permits those flows to be synchronized, making it possible for the whole chain – from supplier through to customer – to operate in step. The result is the process-enabled supply chain.

In this, as some of the case studies in this book show, work is routed quickly and efficiently from one stage in the process to the next. Process times are reduced dramatically, resulting in cost savings in the operations at each part of the supply chain. Time to market can also be dramatically improved, in both developing new products and in shipping products ordered. As a result, customer service and satisfaction, and profit margins, improve. End-to-end monitoring and in-flight corrections all become practical possibilities.

Manufacturers can then integrate product design, production, distribution and logistics into a unified entity. They can weave operating divisions, outside suppliers, distributors and customers into a single, strategically orchestrated whole. There is a dissolution of the operating barriers between organizations.

Many trading organizations already communicate over standardized communications channels, using standardized or commonly agreed data formats and methods for trading. These are part of the trading infrastructure. When they use BPM to control their shared supply chain, they in effect add a process layer to that infrastructure, as depicted below.

At the top of figure 5.7 is the trading trio we introduced in figure 5.5, the firms linked by their respective supply and demand chains. Underlying these is the infrastructure needed to allow them to transact business.

At the bottom is the commercial infrastructure, comprising online markets, electronic payment methods, logistics and the like. Next is the network infrastructure, which includes the participating companies' computers and networks, those of its services suppliers and the Internet itself. The software infrastructure, which we have discussed at length earlier, occupies the middle layer.

Next to the top is the process infrastructure. This layer provides services to the component layers below it by handling the information in the final layer, at the top. It offers a common system for moving information up and down a supply chain. Moreover, it allows those flows to be synchronized, making it possible for the whole chain – from supplier through to customer – to operate in a united fashion.

This sort of vision, which might have seemed futuristic only a few years ago, is now not only practicable, it is in operation in several companies and countries.

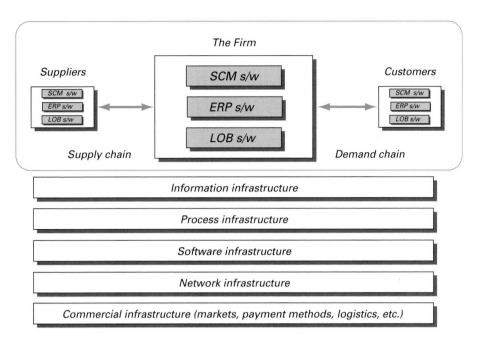

Figure 5.7 Process-enabled infrastructure

Cisco, Hewlett-Packard and General Electric are some of the leading exponents of this unified approach to supply chain management. It is no coincidence that they are all also committed users of BPM.

To show how this might work, here again is our imaginary retailer, ML&S. You might recall that, in chapter 4, we analysed its problems in terms of Stafford Beer's Viable Systems Model. We concluded that ML&S' delivery manager was suffering from a lack of timeliness and frequency in the information supplied to him and that this was possibly low in relevance as well. These deficiencies made it impossible for him to apply suitable corrective action to the problems he was dealing with. The monitoring (sub-system 3a) and control (sub-system 3) within this retailer were nowhere near good enough to permit appropriate actions (sub-system 1) or their coordination (sub-system 2).

Now imagine that ML&S' directors have invested in a thorough and well-implemented programme of BPM. They have also persuaded their trading partners to do likewise. The directors correctly reasoned that, although the problems were mainly theirs, deficiencies in the whole chain needed to be dealt with if efficient end-to-end working were to be possible. (Cisco, Dell, Tesco, Hewlett-Packard and many others share this belief.)

Here once again (figure 5.8) is the diagram of the connections in ML&S' supply chain. This time, we have shown on it the areas of coverage of their and their partners' BPM systems.

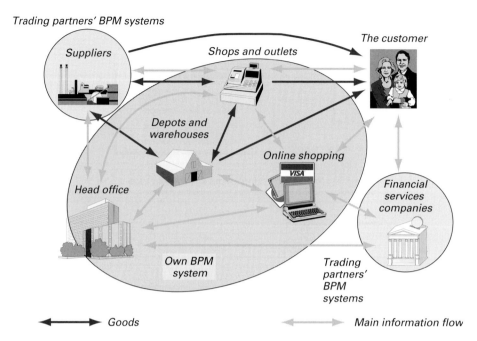

Trading partners' BPM systems

Figure 5.8 Process-enabling ML&S

Within the internal chain, the system (or systems) allows close control and reporting on the movement of goods between ML&S' depots and retail outlets and also out to the customer. Electronic terminals used by delivery drivers permit the wireless transmission of delivery status within minutes. The supply of other internal data, such as Web site transaction details and statistics, arrive as quickly. So, too, does data from suppliers and finance companies, and out to them.

Although ML&S cannot control its trading partners' BPM systems, it can swap process management instructions with them, synchronizing the working of the whole chain. In Viable Systems Model terms, ML&S' monitoring and control sub-systems now permit the triggering of appropriate actions and their coordination.

There are limits to this sharing. As we mentioned in chapter 3, organizations are understandably loath to share their intellectual property by exposing their complete internal processes to potential competitors. They can keep these secrets to themselves by interposing intermediate processes. These expose only interfaces to the full processes, so they become black boxes to those other processes.

It is not just the manufacturing and retail industries that can benefit from process-enabling the supply chain. For some years, banks and insurance companies have pursued the notion of straight-through processing (STP). The idea behind this is to automate the end-to-end processing of transactions, from initiation to resolution. STP aims to make trade processing as automated as possible, allowing business processes to be carried out without unnecessary human intervention.[9] Organizations thus hope to minimize overall lead times, costs and related risks, including inevitable human errors. STP projects are obvious candidates for rapid or eventual inclusion in any programme of BPM.

The desire to achieve the management of end-to-end processes propels a series of programmes within national governments. In Britain, for instance, the government has put in place an active series of projects and expectations directed at what it terms 'joined-up government'. In 2004, it also created the Enterprise Workflow National Project, to help educate local government organizations in the benefits and implications of process management. In the USA, successive Paperwork Reduction Acts, the E-Government Act of 2002 and more recent e-government proposals are going down a similar path.

9 British and Commonwealth readers of a certain age might remember the boast of the sugar refiner, Tate & Lyle, that its product was 'untouched by human hand'. That sums up the main objective of STP.

Case Study 4 Matáv (The Hungarian Telecommunications Company)

Matáv is the main telecoms supplier for Hungary, serving over 3 million customers. Meeting their needs requires 200 technical and administrative processes, some of which demand action within days if not hours. This was not being achieved. Processes were disjointed, paper-based and often unmanageable. Despite being in an IT industry, Matáv made relatively little use of computers to help it run and manage these processes.

The company introduced a process management system from a supplier experienced in the telecoms sector. With the aid of a consultant, this company also handled design, programming and implementation. All this was carried out and controlled centrally. Despite being a company-wide project, directly affecting 1,200 employees, there was little user involvement reported. The most significant implementation issue listed is technical and there are now 'no places for improvisation'.

Perhaps such an approach is typical of the sector, the country or the situation (or all three). It has clearly worked, since Matáv reports an impressive list of improvements. The time taken to introduce new products and services has shrunk from months to weeks. Customers now have detailed information supplied quickly. Users are 'forming queues' to ask for new features and the system is being extended and expanded.

Industry/Sector	Telecommunications	**Location(s)**	Hungary
Group annual turnover/income	2.27 billion euros in 2002 (US$2.75 billion)	**Number of employees in group**	Over 16,000 in 2002
Type of system	Workflow	**Supplier and product**	Fornax WFMS
Number of users	1,200 direct	**Time to complete**	8 months for first; 13 months overall

Business objectives	Improve corporate performance and customer satisfaction
Quantitative results	• 15 per cent increase in efficiency in 2 years • Customer retention capacity increased by more than 10 per cent
Qualitative results	• Faster process execution • Increased individual and corporate efficiency • Reduced revenue loss • Improved network availability • Quick and detailed information available to customers • Tools and data for continuous process analysis and optimization • Further possibilities for process automation

Business background

The Hungarian Telecommunications Company – Matáv (for Magyar Tavkozlesi) – is the dominant supplier in the Hungarian market, providing the full range of telecommunication services throughout the country. About 4,000 people work there on technical processes, including fault repair, network provisioning and internal support and maintenance. This involves 200 business processes, which handle over 15,000 daily requests ('tickets') from the company's 3.2 million customers. The workload equates to more than 100,000 tasks a day.

The company had several problems:

- There were fifty-two fault repair and network provisioning centres, mostly working independently from each other. These had separate coding systems, different processes and individual IT support (or none at all).
- Workforce managers had no computer systems and relied mostly on printouts, telephone calls and fax messages. Sub-contractors could not be directly involved in company processes, nor tracked task-by-task.
- There were no tools available to measure company or individual performance, or to analyse process bottlenecks and other problematical areas.
- The available (and basic) statistical data for different organizational units could not be compared.
- Introducing new services took a long time; IT support for new technologies and processes depended on costly development and training.
- Front office and back office were separated from each other. The back office had limited customer information, while the front office had no information available about current technical processes.

To overcome these difficulties, corporate managers decided to introduce an integrated workflow system to standardize company processes. They wanted improvements in customer satisfaction, individual and corporate performance, and process effectiveness and efficiency.

They set the following goals for the project:

- Improve the quality of fault repair and network provisioning to customers. This had to include shorter execution time, more detailed and more precise information, and greater attention to individual requests and circumstances.
- Reduce the cost of all technical services.
- Document and standardize all technical processes; managing them through one set of rules.
- Make individual and company process performance measurable and comparable. This should create a basis for continuous analysis and improvement.
- Bring new services to market quicker.

System description

In 1999, Matáv selected Fornax to introduce its Workflow and Workforce Management System (WFMS) nationally. System modules have been added since then, to provide new features or extend existing features to new areas of the company's operations. These features include a range of dispatching methods (including SMS text by mobile phone and email), prediction of schedule conflicts, escalations and geographical modelling with travel time calculation.

The WFMS system is continually available and runs on Silicon Graphics Unix servers, over a network set up solely for the workflow system. The servers also run Oracle databases and IBM MQSeries middleware. There are over 1,000 workstations, which use various Microsoft Windows operating systems (9x, NT, 2000 and CE). WFMS connects with eight other major systems within Matáv. The project included integrating handheld computers, all running Windows CE. Matáv intends every technician to have a handheld computer, to further improve process performance.

WFMS has the industry-specific and integration tools needed by a large telecoms operator. These include close integration with general applications, such as those for the call centre, CRM and billing. It also links to the company's network testing system, technical inventories and network management system.

The most important telecoms module is the 'Service Designer'. This is a fully integrated graphical tool for designing, maintaining and displaying telecoms services inside the workflow system itself.

Matáv assigned all fault repair and provisioning responsibilities to three central directorates, These now control the operation of smaller centres. This has allowed the company to use a unified coding convention. This is the basis both of centrally managed business processes and of enterprise-wide statistical analysis.

All business processes are now designed at a high level and rigorously tracked down to the level of elementary tasks and individual executors. Corporate managers can decide whether to specify business rules centrally or leave them for regional managers to define.

Every piece of work is done according to these central or regional rules, the system keeping an account of employees' working time. Some processes run automatically from beginning to end, making them cheaper and quicker.

There is self-upgrading client software for users. When a user launches the software, he is notified if a new client version is available from the server. After a pre-defined period, users are obliged to upgrade the software.

Matáv operates in a region of Europe with multiple languages. WFMS enables switching between an unlimited number of languages on the fly. This applies to display elements, database content and process-dependent information.

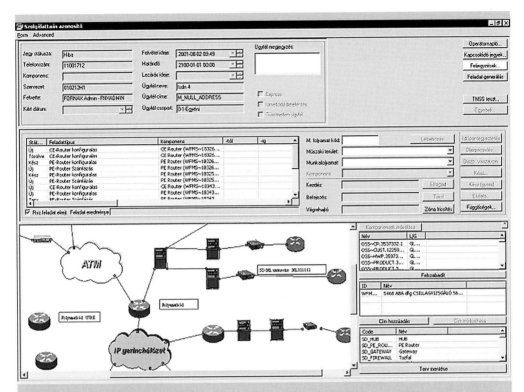

Figure C4.1 Main ticket form in WFMS

WFMS is now the single source of information for all the areas it covers. Call-centre operators, dispatchers and technicians all access the system, directly or indirectly, to enter and read any information related to technical processes.

The screenshot in figure C4.1 shows the main ticket form of WFMS. The upper part of the screen has basic ticket information, such as status, submission date, due date and customer details. In the middle, there is a list of tasks already completed and still under execution. The diagram shows all the relevant technical information inside the workflow system itself. Here, it is the process for a complex data communication network.

Implementation experience

Where and when did the project or system originate?	Centrally, 1999
How long did implementation take?	13 months
Who did the implementing (own staff, contractors, consultants, etc.)?	Fornax

How much bespoke development was there?	Roughly 60 per cent
Were there any special infrastructure needs?	No
What were the most significant implementation issues and how were they dealt with?	Integration to Matáv's monolithic 'legacy' systems running on Tandem hardware including order management, technical inventory and billing systems called OSS; both sides developed a DCE/RPC connector
What lessons were learned?	All the aspects (that is, business process redesign, reorganization, training, internal public relations (PR), system implementation, etc.) of such a project have to be handled with the same care: do not focus on just the system itself
Who is responsible for the system overall?	In Matáv, a separated organization manages the whole application
How was and is training handled?	Training the trainer
What was and is done to encourage use?	Nowadays no further encouragement is necessary; the user community is 'forming queues' in order to submit new feature requests; in the beginning of the implementation, after the first live version of the application a software ergonomic survey was made, to fine-tune the application: managers were convinced by regular management reports

The workflow project handled by Fornax included everything from requirements analysis and system design to development, integration, training and implementation. This was completed by the end of 2000, with almost half of the company using or being managed by the system. The entire project took thirteen months to finish.

Matáv believes that the key to success was the use of a RAD approach. Without it, it feels, the project could not have been carried out in such a short time. One result has been that half the employees of a 10,000-strong company have switched from fax, telephone and MS Excel-based process management, with limited database support.

Strong centralization was achieved. Centrally managed system parameters force users and workgroups always to follow specified business rules. There are ways to overcome unforeseen situations but most processes go in the pre-defined way.

There was some resistance to organizational change and even more to centralization. Fortunately, the project had the full support of regional directors and mid-level managers, so initial problems could be quickly dealt with.

A bigger issue was the introduction of a closed-loop, measurable system. Understandably, people didn't like being tracked all the time and receiving more tasks. Line managers handled these questions, in parallel with system implementation.

User comfort was a major concern. To help users of differing experience, the same functions are available in different ways. An independent consultant carried out a full workplace and software ergonomics survey after the pilot phase, and most of his suggestions have been acted on.

There were three main technical problems that were solved as part of creating the workflow system:

- The Fornax general-purpose workflow management software needed to be converted to something specific to the telecommunications industry.
- The system had to be made tolerant of catastrophes. This was achieved through changes in the server and network architecture and by devising innovative replication methods.
- Change management and configuration management methods were needed. These overcame difficulties in interfacing between the various systems linked to the workflow software.

Benefits and user reaction

What has been the reaction of managers and staff?	The staff at first tried to find arguments why the new system was less efficient than the old one, intending to hide their weaknesses in this way. It took roughly a month using the powerful reporting capabilities of the system to convince everybody that the system was efficient. The management supported the system from the very beginning.
What has been the reaction of customers or trading partners?	Customer dealings were improved though the extensive contact history and automatic call-back capabilities of the system. It increased the feeling of quality.
What has been the overall cost of the system?	Not available.
What have been the main process benefits?	There was a general efficiency improvement. Critical parts of the process can be discovered and the improvement efforts concentrated on these areas.
What have been the main effects on operating style and methods?	The processes are more controlled, with no places for improvisation. Exceptions are handled in a controlled way. The level of the automation increased significantly, in some cases up to 100 per cent. The interaction with customers is more controlled and quality assured.

The results of introducing this new system include:

- faster process execution
- increased individual and corporate efficiency
- quick and detailed information available to customers all the time

- tools and data for continuous process analysis and optimization
- further possibilities for process automation.

The project returned all invested effort in terms of technology, service quality and company performance, while achieving significant cost savings.

There are several factors that together led to excellent financial results:

- Reducing the time needed to clear faults and carry out provisioning tasks minimized lost revenue
- Network availability was improved and operational 'up time' increased
- Improvements in the quality of service increased customer retention by at least 10 per cent; at the same time, customers are now better informed.

Headcount savings of 15 per cent were achieved, including an avoided headcount increase. This came partly from increased technician and dispatcher productivity. The rest came from consolidating smaller organizational units. This saved management and administrative workforce, IT and telecommunication support staff, and equipment and direct operational costs.

Among the intangible benefits of the system are:

- the ability to use and coordinate the work and workforces better for network operations
- consistent, timely and meaningful measurements of quality of service
- the ability to establish an 'outside plant' analysis program to focus on the construction activity, through using standardized work processes and codes
- faster introduction of new products and services, falling from 4–6 months to 2–6 weeks
- standardized and improved management and control for all processes, through consolidating the staff support function.

System users' daily work became simpler. They also now have access to more information, allowing them to follow the entire process from beginning to end. In many cases, people have stopped using other applications.

Matáv's system administrators can now analyse and configure company processes without the help of consultants or programmers. Besides financial advantages, this shortens significantly the time-to-market of new telecoms services. When the company launched its ADSL service package, the IT support for it was provided within 10 days.

The closed-loop workflow system eradicated all kinds of former improvisation. Every event is now done according to centrally specified rules, and is logged and analysed down to the smallest detail. One of the main achievements is the optimized scheduling of resources. This is true both for on-the-field technicians and also for application systems that are connected to WFMS through an interface.

Between them, Matáv and Fornax have created a tool for controlling and streamlining the main technical processes of the organization. By the end of the project, an underdeveloped part of the company had become the leading driver from both the technical and the business point of view.

The future

There are several plans for the workflow system. One objective is to extend it to more applications, building more interfaces to and from WFMS. Priority will go to those projects where complete process automation can be achieved.

There are also plans to extend workflow and workforce management beyond the technical processes. Matáv is considering applying the existing system features to almost the whole company.

The company feels that there is huge potential in handheld computers. The existing pilot will be widened to other areas. It expects further increases of company performance from this, as well as significant headcount savings.

6 Organizations, people and systems

First thoughts

As we said in chapter 1, this is not a cookbook. Cookbooks preach one way of making dishes, to their authors' favourite recipes. They assume that you have all his preferred ingredients to hand, a full range of utensils and the time to follow his methods. All this is nice in theory but seldom possible in practice.

To continue the analogy, our aim instead is to help you become a good cook. A good cook can make something nutritious and attractive whatever ingredients and tools are available. He can adapt himself and his methods to the circumstances he finds himself in. A good cook still uses cookery books, but as a guide, not a prop.

So it is with managing processes. You are unlikely to be in an ideal organization, with ideal people, ideal computer systems, an ideal business strategy and an ideal management style – however you might wish to define 'ideal'. As with being a good cook, managing organizations is a matter of making the best with what you have. That is the basis on which we have written this book, this chapter especially.

We are not believers in 'one best way' except perhaps in closely defined technical matters. With organizations, even if such a thing were possible in theory, how could one prove that any measure was in fact the best? Organizations are not laboratories and there is no control group with which to compare results. Much as you may sometimes wish, you cannot create a set of identical replicas of, say, the marketing function and subject each to different conditions to see how it reacts.

Even if there were a single best way possible we could not tell you at a distance what that is for your own organization and circumstances. How can we, since we know nothing of you or them? All we can safely say is that compromises with a so-called ideal will be inevitable, so we favour pragmatism over evangelism.

It should, though, be an enlightened pragmatism. If you have a fuller idea of what is possible as you make your decisions and agreements, the less likely it is that you and others will regret them later. In this chapter, therefore, we look at some of the main dimensions of the way people influence processes and vice versa. We set out some principles that, we hope, will help you decide on the approaches, methods and tools that your organization will use. We feel these will increase your chances of producing good results in the long and short term.

The ground we cover

In the introduction to the third edition of his book, *Understanding Organizations*, Charles Handy says this:

I came to the study of people in organizations expecting certainty and absolute knowledge in the behavioural sciences. I anticipated that I would find laws governing the behaviour of people and of organizations as sure and as immutable as the laws of the physical sciences. I was disappointed.

Handy goes on to say that there were two good reasons why that predictive certainty is not possible in studying people in organizations. The first is that there are too many variables and too little data on them to make predictions (his book deals with sixty of these variables.) The other reason he cites is that people often act in ways that override many of these influences on their behaviour.

We have not space to cover all the variables that Handy considers. We therefore look briefly in this chapter at thirteen of them. Figure 6.1 shows an

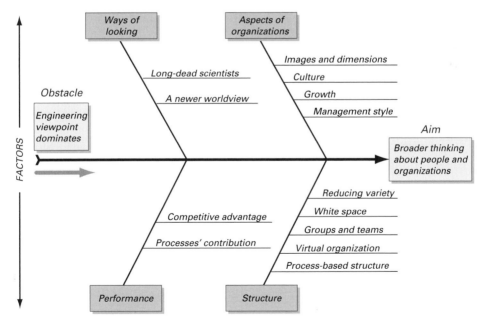

Figure 6.1 Topics we cover

arrangement of them in an Ishikawa diagram, a widely used tool for problem elucidation.[1]

If you are familiar with the study of organizations, you will see that we have omitted several topics of undoubted importance. Matters such as job satisfaction, motivation, leadership, creativity, change management and recruitment do not appear in this chapter. This is not because we think them irrelevant but solely because there is not space to discuss them. You should consult other sources on these matters. We concentrate instead on those factors that we feel most directly affect and are affected by the computerization of business processes.

Ways of looking at organizations

The dead hand of long-dead scientists

Computer systems analysis and programming are engineering disciplines at heart. So, too, are industrial engineering (naturally enough), methods study, quality assurance (QA) and much management consulting. Engineers have a particular view of the world, which has been of immeasurable value in solving engineering problems. It is, alas, not much use elsewhere and is often a hindrance. The problem for and with engineers is that they often do not know this. Instead, they usually try to apply their approach to all problems and situations. Two of your authors are engineers, so are familiar with this habit of mind.

William E. Kastenberg, another engineer, describes it here, in an 2003 article from *Science and Engineering Ethics:*

For nearly forty years, [I have] been training engineers to see the world by breaking it down into its component parts (atomistic), to frame technical problems in terms of cause and effect relationships (deterministic), and to seek the solution to these problems without considering the possible influence of the observer on his perception of the observed world (dualistic). This way of understanding the world is characteristic of the linear or reductionist paradigm that has dominated Western thinking for almost 300 years and underlies the current culture of engineering.

From the start of the Industrial Revolution, through to Taylorism, Henri Fayol and beyond, work management has been based on the division of labour. This is an engineering approach, which breaks work down into its component parts and treats each part separately. Frederick Winslow Taylor's ideas about 'scientific management' epitomise this.

---

1 This is also called a fishbone diagram, cause-and-effect diagram and root cause analysis. It is commonly used in quality improvement, by individuals and groups, and is named for its inventor, Kaoru Ishikawa. He popularized the technique in the 1960s, while working at the Kawasaki shipyards in Japan.

The irony is that these ideas, even while they were being formulated, were based on outmoded scientific thinking. They embodied the worldviews of people such as Isaac Newton, René Descartes and Francis Bacon, ideas that were constraining and inadequate even by the end of the nineteenth century. At the time Taylor was promoting his methods – and Henry Ford was making the first Model T car – scientists such as Max Planck, Albert Einstein and Nils Bohr were revolutionising the way we look at the world. This was around a hundred years ago but people are only just beginning to absorb these pioneers' work.[2]

Sir Geoffrey Vickers reinforced the point in 1984 in his book, *Human Systems are Different*:

It is sobering to reflect that even three or four decades ago it was still possible for intelligent people to debate whether a 'whole' could possibly be more than the sum of its 'parts', and if so how. Today it is perfectly apparent, though not yet completely accepted, that every whole is bound to be both more and less than the sum of its parts. It is less because its parts are constrained by being organized. They can no longer do some things which were open to them in their unorganized state. For example, the constituents of a dead body are free to combine in other forms, once the discipline of the living organism is relaxed. [See the apple tree example in chapter 1.] But they are also more than their sum because, when organized, they can do what they could not do alone or as an unorganized aggregate.

A newer worldview

The scientist Fritjof Capra summarized this 'new' thinking as being based on five principles, which we paraphrase below:
1. Look at the whole not the parts. The properties of the parts can be understood only from the dynamics of the whole.
2. Look at the process not the structure. Every structure is a manifestation of an underlying process.
3. Do not treat data as objective. Who gathers it, why and how all change the nature of evidence. Make clear how you found out the 'facts' you present and not just the facts themselves.
4. Use networks not buildings as the metaphor. Connection and flow replace structures and fixity.
5. Recognize that descriptions are imprecise and do not state an absolute and unchanging truth. All scientific concepts and theories are limited and approximate; none is complete and exact.

There is a sixth principle we would add, which is a corollary to these – the presence of an observer changes the system. This also is a perception from early

---

2 Tom Peters, the management writer, makes the further point that science proceeds by making mistakes, then having them publicly recognized and corrected. Organizations practising so-called Scientific Management typically go in the other direction, internally punishing mistakes and correcting them in secret.

twentieth-century science.[3] It found a humanistic parallel in Elton Mayo's work at Western Electric's Hawthorne factory in the 1920s and 1930s. The results appeared to show that watching how work was done itself produces an improvement in productivity. This has since become known as the 'Hawthorne effect', and has been used as a sort of social placebo by manipulative managers.[4]

What is less disputable is that observers are seldom, if ever objective. Like everyone else, they have values, beliefs, norms and motives that colour their perceptions. Nor are they all-seeing. What they do see is filtered through their personal perceptual apparatus, with all its biases and defects. This applies also to where they choose to draw the boundaries of a system. There is nothing anyone can do to stop these subjective influences; recognizing that they exist is a step forward.

Some of this might seem radical and abstruse to you but it shows the way things are moving. It does not mean we jettison all that the past has taught us – Newton's Laws of Motions still apply in everyday life, for example. Nor does it mean we have gone New Age on you. That involves a rejection of science.

What it does mean is that we should recognize the way these old ideas mould our thinking. As the writer and educator Alfie Kohn says:

There is a time to admire the grace and persuasive power of an idea; there is a time to fear its hold over us. The time to worry is when the idea is so widely shared that we no longer notice it, when it is so deeply rooted that it feels to us like plain common sense. At the point when objections are not answered any more because they are no longer raised, we are not in control: we do not have the idea; it has us.

Our aim throughout this book is to help you think afresh about business processes and their management. There are two impediments to this:
- The models and ideas that you already have in your head
- Those models and ideas that BPM suppliers and consultants will try to put there.

Without exception, everybody carries sets of ideas in his mind of how things ought to be in business and organizations ... and life. People are usually unaware of how they arrived at these models and, often, that they even have them. These

---

3 Max Planck formulated his quantum theory in December 1900, although its implications for the role of the observer needed Nils Bohr and Werner Heizenberg to elucidate them in the 1920s and 1930s.

4 Mayo and his colleagues reported several outcomes of their varied researches, any of which might justifiably be labelled the 'Hawthorne effect'. For example, they also reported that workgroup norms influenced productivity more than individual abilities and that the social aspects of work were as important to people as the economic returns. There has been much subsequent debate over the validity of their investigations. Perhaps the only consensus is that trying to ascribe simple, single causes to social effects is bound to fail. The main references are Elton Mayo's *The Human Problems of an Industrial Civilization* and Fritz Roethlisberger and William Dickson's *Management and the Worker*.

unacknowledged notions – heard for example in expressions such as 'Everybody knows that . . .' – get in the way of learning.

There is a famous quotation about this from John Maynard Keynes:

Practical men, who believe themselves to be quite exempt from any intellectual influences, are usually the slave of some defunct economist. Madmen in authority, who hear voices in the air, are distilling their frenzy from some academic scribbler of a few years back. I am sure that the power of vested interests is vastly exaggerated compared with the gradual encroachment of ideas.

One of our objectives in this chapter is to drown out the voices of BPM's 'defunct economists'. If you are not already doing so, we want you to start thinking afresh about business processes and systems and how they should be managed. BPM suppliers and consultants are people, too. They each also have their unknown and undeclared models. In addition, they will have the models and ideas that are embodied in their products and methods. You will find that they often rush into descriptions of their offerings without declaring the model of processes and business underlying them. We hope that, after reading this book, you will feel confident enough to challenge them to make that model clear. How else can you assess its value to you?

Trends in thinking about organizations

There are more ways of thinking about organizations and management than the two we discuss above. We have cited these because they are so dissimilar. It is conventional to divide them into four main approaches, or schools of thought – 'classical' (including Scientific Management), human relations, systems and contingent. Table 6.1 set outs the main features of each, with the names of some of the main contributors to the approach.

Consciously or not, different managers apply different approaches and differing combinations of them as circumstances vary. In this book, we mainly follow a contingent model.

A table like table 6.1 is, we hope, useful as a guide to the main currents of thinking. The problem is that its format implies sharp differences that do not exist in reality. It also suggests similarities between adherents of these schools that may not exist. Further, you may disagree about whom we have placed where in the 'influential thinkers' column.

None of these difficulties is particular to this table; they are true of any such summary of a complex reality. As we point out elsewhere, it is a model and therefore a simplification.

Table 6.1 Some approaches to organizational thinking

Approach	Main ideas	Strengths	Weaknesses	Influential thinkers
Classical	Uniformity, commonality, and control Emphasizes formalized and specialized roles and functions, arranged in a hierarchy	Orderly, rational, disciplined and predictable Highlights structure and formal rules Strong on repetitive processes and individual productivity	Mechanistic, impersonal and inflexible Emphasizes vertical thinking Largely ignores the influence of groups and external changes Undervalues individual needs and wants Encourages rule-following rather than objective seeking	Adam Smith, Max Weber, Henri Fayol, F. W. Taylor, Frank and Lilian Gilbreth, Henry Ford
Human relations	Sees organizations as social arrangements Emphasizes motivation, morale and job satisfaction Brings in consideration of conflict and of leadership and management styles	Highlights the influence and variability of teams and of individuals' needs and wants Takes account of the informal organization Encourages employee participation in decisions and recognizes its importance	Can degenerate into woolly 'do-gooding' and excessive introspection Sometimes seen (sometimes justifiably) as manipulative, not changing underlying power disparities	Robert Owen, Elton Mayo, Mary Parker Follett, Peter Drucker, Abraham Maslow, Douglas McGregor, Chris Argyris, Joseph Juran, W. Edwards Deming
Systems	Sees organizations as self-sustaining organisms, influenced by and influencing their environment Reductionist version is the basis of management accounting and most corporate computing	Emphasizes connectedness, timing, learning and modelling Encourages 'end-to-end' thinking	Not always fully understood, especially at higher organizational levels Accounting and IT versions usually overvalue logic and predictability and underestimate human variability and contribution	Ludwig van Bertalanffy, Stafford Beer, Herbert Simon, Michael Hammer, Michael Porter, Peter Senge
Contingency	There is no one best way suitable for all situations Acknowledges the value of and builds on the other viewpoints. The basis of BPM	Can lead to the closest match between problems and solutions, and between aims and methods Increases managers' ability to cope with variety (see later)	Needs open-mindedness and broad knowledge from managers Can degenerate into a 'pick-and-mix' approach, lacking rigour Danger of assuming that a good 'fit' among factors equates to good performance	Fons Trompenaars, Joan Woodward, Roger Harrison, Charles Handy, Ricardo Semler

Images of organizations

Gareth Morgan made a more detailed analysis in the 1980s. He listed eight main ways that people think about organizations:

- as machines
- as organisms
- as brains (as in the Viable System Model)
- as cultures
- as political systems
- as 'mental prisons'
- as 'flux and transformation'
- as 'instruments of domination'.

All these are self-explanatory, except perhaps 'mental prisons'. This refers to the illusions people create for themselves when thinking about situations. The self-reinforcing nature of these illusions is illustrated in the 'ladder of inference', an analogy devised by Chris Argyris (figure 6.2).

Unless people change or at least test their perceptions, values and assumptions as they learn, they will continue seeing the world the same way. Argyris calls this 'single-loop learning'. (He feels that TQM, for example, typifies this.)

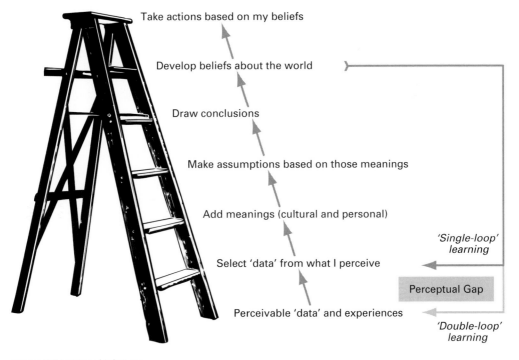

Figure 6.2 Ladder of inference

'Double-loop' learning takes place when people question the way they arrive at their judgements and decisions, and when they see the selective nature of their data-gathering. At this stage, they start to close the perceptual gap shown on the diagram. They are still being selective, but knowingly so, aware of what they are leaving aside.

The models that Morgan describes are neither wholly right nor wholly wrong. Like the schools of thought listed in table 6.1, each makes a useful but partial contribution to the overall picture. They are all, in the logician's phrase, necessary but not sufficient.

There is a practical lesson here: that no single view of organizations should be allowed to dominate. If it is, other and equally useful viewpoints can then easily be dismissed as invalid. This, in turn, invalidates the opinions and expectations of people who see the world that way. If it is to succeed, BPM cannot afford to disenfranchise any group of people.

Dimensions of organizational behaviour

Figure 6.3 and table 6.2 set out and describe some the main factors that, in combination, help make the difference between one company and another.

All the eight elements shown have an effect on and are affected by computer systems. The lower three elements in the diagram – *Structure*, *Technology* and *Physical Setting* – are the deterministic or 'hard' factors. These are the ones that can be directly

Figure 6.3 Factors in organizational character

Table 6.2 Organizational factors explained and explored

Factor	Meaning	Some system design considerations
Purpose	What the company is there for, what its objectives are. What its 'message to the world' is	Do systems designers and other employees and the know these? Are they embodied in the systems strategy? Do people feel they should be?
Group relations	An aspect of culture How different groups and functions feel about each other How they deal with each other	Will formality increase or decrease? Will users and non-users drift apart? Will group friction interfere with system operation? Will any group or groups hoard information?
Communications flow	How information moves around the organization, and into and out of it Covers word of mouth and paper, as well as electronic media	Will more communication equal better communication? Will electronic methods be too fast for some people or groups? Will they by-pass some useful non-electronic channels? Will everyone need or want electronic methods?
Structure	The 'formal organization', as shown on the chart on the wall	Will traditional functions be retained? Can IT assist or enable structural change? Where will systems responsibility reside?
Technology	All systems and technologies, not just IT Includes those of product design, production and distribution Also project management, personnel, budgeting and accounting systems (Paper-based systems are technologies, too)	How familiar with new technology is the company? Is a pioneer, rationalist or laggard? What is the predominant system design approach or philosophy? How quickly is payback expected? Who 'owns' the systems – IT, users or both?

Table 6.2 (*cont.*)

Factor	Meaning	Some system design considerations
Physical setting	Both the immediate and overall physical environment in which the company and its employees operate Covers all aspects from the choice of desk to the design of offices and sites, and their location	Are the present buildings needed for the work proposed? Is relocation likely or planned? Is 'hot desking', remote working or home working planned? Does the company operate from one site or several? Does it operate internationally?
Reward systems	How the organization shows people it thinks well of them Covers more than pay – includes bonuses, 'perks', status symbols, public recognition and promotion Also includes measures and controls	Will the new system reward relevant behaviour? Will machine monitoring be used? What will be the effects on workloads, hours, pay, holidays and careers? How will the company deal with those who cannot cope with the new system(s)?
Rules and traditions	The main aspect of corporate culture The explicit rules and, mainly, the unwritten ones ('how we do things round here')	Has any one person the power and the will to try to change them? Will the new systems run counter to the corporate culture? Is the design team aware of the prevailing culture? Is there just one culture to be aware of?

controlled. They are the elements that are open to quantified analysis and to systematic investigation. Typically, they are the main focus of attention in computerization.

The upper three elements – '*Rules' and Traditions*, *Purpose* and *Group Relations* – are probabilistic or 'soft' aspects. They are equivalent to the human body's autonomic systems (the heartbeat, sweating and so on). Like them, they are controllable only indirectly at best.

In the same way as a person's resting pulse cannot be altered in the short term (except, perhaps, by skilled yogis), neither can these aspects of organizational life. Only longer-term attention to diet, exercise and way of life make a positive difference to the fitness of the human heart. Only longer-term attention to leadership, trust, consistency and openness have a similarly beneficial effect on the workings of the body corporate. (There is more on culture in the next section.)

Between these two sets of factors lie *Reward Systems* and *Communications Flow*. These appear midway, on the horizontal axis, because they contain a combination of 'hard' and 'soft' elements. Reward systems, for example, comprise directly controllable aspects such as pay and measure of performance. They also contain indirect matters such as respect and recognition. Similarly, communications flow deals with both formal and informal information channels.

Western business culture has traditionally emphasized the deterministic elements, the lower half of figure 6.3. The others are sometimes even sneered at. They are derided as being 'soft', unscientific and feminine (as if this were a weakness). Yet in other parts of the world, notably Japan, businesses consciously seek to integrate all the elements in their operations.

National and ethnic cultures play a part as well. What works in Michigan will not necessarily work in Essen, Milan, Tokyo or the Punjab.[5] Even within the same Asian country, for instance, something that appeals to, say, European-educated meritocrats might not sit well with locally educated workers.

Maintaining balance

If one set of these factors is radically changed but not the others, the result is organizational and individual stress. The other elements have to change to keep step. It is as likely, if not more so, that they will resist the proposed change.

Some organizational factors can be changed quickly at low risk. Technology is an example, perhaps by introducing workflow automation. As the case studies show, this can be accomplished in weeks or months.

If the change is congruent with the other, more deep-rooted aspects, there should be few problems. If it is not, these will work to delay or defeat the change. It takes a traumatic or cataclysmic event to produce a fast change in factors such as purpose, group relations or rules and traditions. Usually something like company bankruptcy or takeover, or a brutal de-layering programme, is needed to effect this.

In the normal course of affairs, changing these more deeply rooted elements takes many months, sometimes years. This is why properly executed business change programmes last so long and are so difficult to bring off. Great care is necessary.

Contrary to popular management myth, individual human beings are not resistant to change. They go on foreign holidays, marry (and remarry), move house and make other abrupt shifts in their life's direction with none of the obstructiveness they show at work. What people really resist is imposed change, change that they do not understand, do not have a voice in, do not feel any control over and whose

---

5 Space does not permit a longer discussion of the influence of national cultures on management style. If you wish to explore the topic, an excellent and readable starting point is Fons Trompenaars' 1993 book, *Riding the Waves of Culture: Understanding Cultural Diversity in Business.*

purpose they mistrust. The answer is to help them understand, to give them a say, some control and a reason to trust. If you can, try to get them to lead the change – or, at least, be active participants in it.

When I hear the word 'culture' . . .

. . . I reach for a set of job descriptions? A training plan? An organization chart?

These are mechanistic elements of work, not aspects of culture. They affect culture, and are sometimes an expression of it, but are not in themselves cultural variables. To regard them as such is the dumbing down of an idea.

Equally boneheaded is the idea, beloved of macho managers, that cultural change can be bulldozed though. At best, such programmes alter some of the deterministic elements of the organization's character, possibly violently. The other elements, including the real culture, may or may not change to catch up. Usually these elements end up at odds with the variables that have been 'reengineered', producing tension and strife.

Change is important and often necessary but there are better ways to bring it about. Trying to understand the culture and then working with the grain of it normally produces longer-lasting results. Culture is a powerful force in the organizational as well as the national context. It establishes patterns of expectation, it is the filter through which experiences are interpreted and it is the silent language that binds the organization together.

Organizations differ in their culture. No two organizations are alike and no two parts of the same organization are alike. There are cultures based on people's occupation, cultures based on the business of the organizational unit they work in and cultures based on location. These all mingle with the organizational culture to create the distinctive character of that specific unit.

Any system or system architecture that runs counter to the cultural norms of an organization is, at best, likely to be hard to implement in a lasting way. It will more likely fall into disuse or be changed by its users into something more consonant with their wishes. In an equal fight between culture and technology, the latter will lose.

Business has its own cultures, so too does computing. To become aware of the nature of those cultures while living in them is an essential first step to truly understanding how they work. The fish need to become aware of the water in which they swim. It is the first step to helping them to change.

The growing organization

Size is an important factor in accounting for the differences between organizations but it is not an overriding one. Many large corporations, such as Kyocera in Japan, are purposely set up and run as though they were collections of small companies.

Conversely, there are numerous middle-sized companies whose directors feel obliged to imitate the arthritic and divisive ways of a 1970s corporate giant. As the case studies show, companies of all sizes and in all industries can make productive use of process management systems.

Another aspect of organizational sensing is growth. Companies are not static. They are dynamic systems with different characteristics at different stages in their existence. A big company is not just a small company with more people; it is a different animal.

One of the best-known models for describing these stages is that by Larry Greiner, which he published in the 1970s.[6] He suggests that there are four main stages – direction, delegation, coordination and collaboration – and describes what happens during each metamorphosis, and what triggers it.

Each growth phase produces a different set of demands on the organization's processes and computer systems. This does not necessarily mean introducing a new system each time. It does, though, mean that the objectives for the system and the way of using it should change each time. An adaptable system is essential for this but, then, designing with change in mind should be a standard practice in modern systems design.

Another implication is that a transitional phase is often not a good time to try introducing any substantial computer system. The necessary management vision, commitment and follow-through will all be lacking. It will also be harder than usual to draw up and test a coherent set of system requirements.

Management styles

Does your organization like people? It's a simple enough question, but one fundamentally important to the way it will go about BPM.

Note that the question is not if your organization *says* it likes people. That is a different matter entirely. Examples abound of corporate utterances about 'our people being our most important resource' or similar. Most of these are simply hyperbolic; many are downright hypocritical. It is the reality – the truth of the matter – that is at issue here.

So, back to the question: does your organization like people? Does it value them, look after them, take their views into account, engage their entire person in their work and generally treat them as unique and creative beings? Or does it deal with them as though they were identical modules of productivity, to be hired and

---

6 See 'Evolution and Revolution as Organizations Grow', *Harvard Business Review*, 1972. As will have become apparent by now, models and descriptions of human, organizational, business or economic behaviour do not date rapidly. They can be valid and useful for decades, sometimes centuries, after first publication. Information about computer products and suppliers is the reverse – it has a shelf life sometimes measured in weeks. Fitting one's mind round the timescales and *tempi* of both can be difficult but it is a useful ability to acquire.

fired at a moment's notice, to be told only those things they need to know and to do just what they are told, when they are told? Some of these matters are an aspect of corporate culture but many of them are the result of the prevailing management style. We have presented a dualistic view here, based on Douglas McGregor's well-known division into 'Theory X' and 'Theory Y', but there are many others.[7]

As elsewhere in this chapter, the main point is to recognize that there *are* different styles of management and of leadership. These play an important part on how systems and projects are selected, funded, installed and managed.

It is also worth remembering that people's management and leadership styles often vary with circumstances. Only rigid managers use the same style everywhere, every time and for everyone. This does not necessarily make them bad managers but it suggests they are not well suited to changing times and new ways of working.

Organizational performance

Sustainable competitive advantage

The origins of this expression are lost in the history of modern management. It is rooted in the belief that one or a few attributes of an organization will guarantee its longevity and success. All one needs to do is identify and isolate these and then buff them up. The inevitable result is that people concentrate on these few characteristics to the exclusion of other important factors.

There have been many candidates put forward as true sources of this elusive quality. They include adaptability, alliances and networks, brand equity, culture, human capital, innovation, intellectual capital, internal communication, leadership, quickness to learn, reputation and strategy execution. Clearly, each commentator has his own favourites.

In reality, organizations are too complex to boil down to a few variables. Their people, purposes, customers, suppliers and partners, traditions, accounting and other control systems and so on all matter. Each is important, each is necessary and all must be kept in balance.

As to sustainability, a sort of answer comes in the book, *The Living Company*, by Arie de Geus. In this, he reports that the life expectancy of the average large

---

7 See *The Human Side of Enterprise*, from 1960. Among other well-known typologies are the seven-part model by Ralph Tannenbaum and Warren Schmidt, and William Ouchi's comparison of American and Japanese ways of managing.

company is a mere forty years. Only sixteen of the 100 largest US companies at the beginning of the 1900s still existed when he wrote the book in 1997.

de Geus identifies four characteristics of long-lived companies. It is a view of the organization as living system:

- They are sensitive to their environment. They sample, learn, and adapt to what is going on around them.
- They have a strong *persona* or sense of identity, based on their ability to build a shared community.
- They are patient and tolerant, allowing diversity around the edges. These peripheral activities may later become part of the core.
- They are frugal with their money, which they use to govern their own growth and to give them options.[8]

We like this analysis but this may simply be a reflection of how we see the world. As with all the models and ideas in this book, it is your choice to accept or reject it, to act on it or pass over it.

Longevity is also a matter of turbulence, or rather the lack of it. Companies such as Deutsche Telekom and some American airlines have long enjoyed a stable operating environment, being protected from competition though regulation or industry practices. Some others, such as General Motors and Nestlé, have had relatively stable product lines. Organizations who do not occupy such lucrative comfort zones must adapt and innovate. Adaptability, which BPM can help bring about, is a primary requisite for survival.

The contribution of processes

Modern organizations are utterly dependent on electronic information. Their telephones, computers, Internet links, data channels and email are their nervous system. Without them, they cannot stay alive for more than a few days. They can rightly be called electronic businesses. These have processes at their heart, along with people and information (figure 6.4).

These three assets help give an organization its soul, its energy and, as discussed above, its competitive edge. Any of them can be stolen or copied but they cannot accurately be replicated. This is makes them unique, especially in combination, and makes them worth guarding.

Your competitors can readily reproduce your other resources. Technology – of design, production, distribution or customer interaction – is easily copied. Any

---

8 Harking back to what we said about observers' biases, this seems to be a thoroughly Dutch model. It is perhaps no coincidence that de Geus worked for many years for Royal Dutch Shell and is himself Dutch. Against that, one can argue that his findings are based on thousands of observations. It is in the perverse nature of such exercises that, at the time of writing, Shell itself should be suffering some financial setbacks.

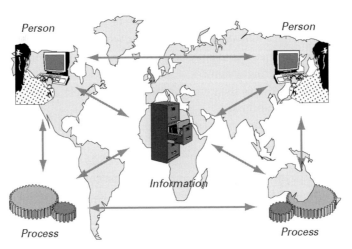

Figure 6.4 Major internal resources

advantage gained from this is usually short-lived, as other people find similar tools or simply buy the same ones from your suppliers.[9]

Financial resources are also easily mimicked. Money is the ultimate commodity, after all. Land, property, goods and stock are also commodities. None of them gives lasting superiority.

With the rise of e-commerce and interorganizational process automation, those resources reach beyond the boundaries of the organization. Your customers, contractors and service suppliers can communicate readily with your employees, information and process. Your 'upstream' trading partners can do so even more intimately, so much so that it can be hard to tell where you end and they start. All this makes optimizing these resources all the more important and worthwhile (figure 6.5).

Assessing the likely effects of BPM on an organization is not a simple matter. Despite what some gung ho consultants might maintain, there is no recipe for success and no certainty about outcomes, for the reasons that Handy gives above. Organizations are social arrangements with an overlay of technical and financial systems (accounting is a technology), and this technological superstructure is sometimes not a good fit on its human foundation.

Think of the organization as a lorry. The whole vehicle is in motion but progress is not always certain or smooth. There are areas of wear and chafing where the body

---

9 Despite its temporary nature, technological advantage is still jealously guarded. As we found when trying to gather the case studies for this book, many organizations are deeply reticent about the systems they use. They would happily be prepared tell us about their objectives, structure, processes and similar but not in detail about the technical means used. Some would not permit their cases to be published (or even republished) for fear of letting competitors know even who supplied those systems. Pointing out that this information would be at least a year old by the time the book appeared made no difference.

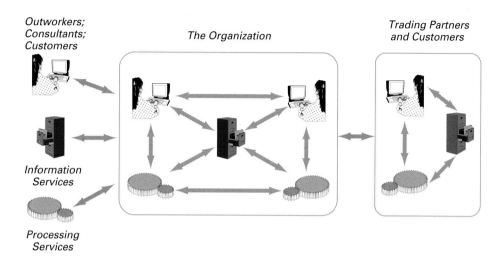

Figure 6.5 Extending the reach of internal resources

meets the chassis, there are gaps where they do not meet and there are stress points all over. These sometimes become points of fracture.

Business process improvement is about making the lorry perform better. BPM, on the other hand, is about doing this *and* improving its steering *and* improving the fit between body and chassis. Moreover, it is about doing all this is while the lorry is in motion; you cannot pull in anywhere to make repairs.

(There are those who say you should instead be turning the lorry into some other kind of vehicle, a bus perhaps, or a helicopter. We discuss those sorts of questions in chapter 8.)

Here, then, are our suggestions. We assume in making them, as we do throughout the book, that you are familiar with the main texts and theories. In case you want to refresh your knowledge of some of them, there is a suggested reading list at the end of the book.

The tyranny of the organization chart

Most managers are like Soviet bureaucrats, living in a dual world – the real world and the world of officially sanctioned ideology. (*The Witch Doctors*, John Micklethwait and Adrian Wooldridge)

Hanging on the wall, printed in the handbook and available on the Intranet is a work of corporate fiction – the organization chart (organogram). We all use it, we all believe in it to some extent and yet we all know it to be misleading. Its defects are well known – it is two-dimensional (usually), it is fixed (if printed) and it is seldom up to date. At best, it presents a picture of one constrained facet of organizational life.

Despite these shortcomings, the organization chart has value. It is particularly useful for portraying administrative arrangements, for example. Online telephone directories, personnel records and security systems rely on it.

From a process management perspective, there are two difficulties with basing the design of computer systems on the organogram. The first, as we have said, is that it is fixed. The other is that it is does not show the flows of work or communication.

Rather, the organogram portrays the 'official' organization. This relies on control, order and predictability. It is a world of job descriptions, reporting lines and budgets. By contrast, the unofficial organization is a place of shortcuts, contacts and dynamic networks. It runs on influence, trust and personal transactions ('favours').

The unofficial organization exists for one reason, which is that, on its own, the official organization seldom works. Even in the most rigorously planned and 'procedurized' business, people have to work around the rules in order to get things done. The more highly ordered the organization, the more important is the work of the unofficial side to its overall effectiveness.

A curious recognition of this importance arises in Britain whenever trade unionists declare their intent to 'work to rule'. This is always recognized for what it is: a threat, not a promise. Its members are saying that, unless their demands are met, they will put only the official organization into operation. Heaven help the employer and its customers while they are doing so.

Reducing variety

In systems terminology, as we discussed in chapter 4, this problem arises because the formal organization does not posses the requisite variety to deal with the unusual. When asked to deal with a matter outside its ability to respond, it either reacts inappropriately or simply ignores the stimulus.[10] Often, and usually without acknowledging or recognizing the fact, the official organization allows the unofficial organization to take over. The organization and its people then muddle through.

Where the formal organization is too strong or rigid, this coping mechanism is prevented from coming to the rescue. The typical result is that, in the words of an old advertisement for insurance, a problem turns into a crisis. Inflexible and out-of-date processes embodied in central computer programs are frequent contributors to such escalating difficulties.

People can suffer from the same weakness. Chris Argyris' ladder in figure 6.2 shows how they habitually cope with excessive variety. An organization does much same. To remedy this you either to try to make its systems able to respond to all

---

10 At the risk of labouring the point, this is one of the reasons why engineering specialists are traditionally poor at dealing with human and organizational issues. These matters often do not appear on their radar.

possible eventualities or else design them to filter out the bulk of them. The former is impossible. A system able to deal with even 95 per cent of possible events would be fiendishly difficult and expensive to create. It also poses the question of how you would know what makes up 95 per cent of possible eventualities or how important the missing 5 per cent might be.

The alternative is to filter them out through some form of variety reduction. Another name for this is 'standardization'. Purchasing departments do this, winnowing out infrequently used suppliers or rarely arising purchases. Nursing departments in hospitals have a routine they call triage. They sort patients by the seriousness of their condition, treating the urgent first.

Similar principles apply in designing organizational systems. You put some apparatus or logic into the system that will group inputs into fewer categories. Sometimes, you also want it to decide which inputs to ignore.

The device that does this best is the human being. Equipped with unrivalled abilities in pattern recognition and decision-making, it is self-correcting, mobile, versatile and able to learn. It is usually also goal-directed rather than blindly rule-following. (The sub-species known as the 'jobsworth' is an exception to this.)

We are only half joking. If you take a 'black box' view of human abilities, only an extraordinarily versatile and powerful computer could match them. It does not yet exist. This is one reason it is so hard to computerize people out of organizational systems.

When you group people into the entity called 'the office' or 'the team', you create an even more versatile sub-system for reducing variety. Offices and workgroups insulate the innermost parts of the organization, those that rely on routine and predictability, from the unruliness and randomness of the business environment. They filter out the exceptional, the unexpected and the unprogrammable.

White space

Sometimes people take the 'black box' view of humans too far. This quotation offers an example.[11]

It's not just the old systems, it's everything that's wrapped around them, including old processes, old work-arounds, and the white spaces where you have to stick clerks into the process because the systems are incomplete.

It seems this person has a low opinion of involving human beings in processes, portraying them as cross between putty and fuse wire. More interestingly, he uses the popular analogy, that of 'white spaces' on the organogram.

11 We have made it anonymous, having no wish to pillory its author.

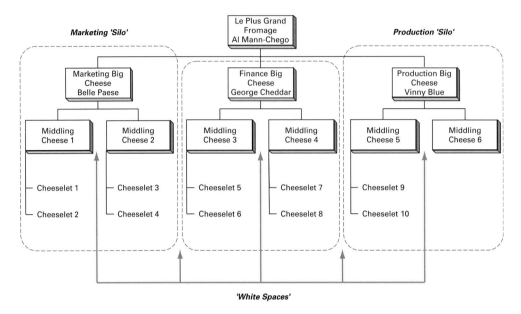

Figure 6.6 Cheese Inc.'s white spaces and silos

As we mentioned in chapter 2, the 'white spaces' expression arose from Rummler and Brache's book, *Improving Performance: Managing the White Space on the Organization Chart*. White space is the gaps and the interfaces between organizational functions.

In figure 6.6, there is a simplified chart of a fictitious organization, Cheese, Inc. Overlaid on it are dashed lines showing where you would expect to find computing 'silos', 'stovepipes' or 'chimneys'. These are computing arrangements that concentrate, sometimes exclusively, on the data and processes within that function. As a result, horizontal computer-based processes crossing between functions are difficult to institute. They are impossible across the entire organization.

At the bottom of the diagram are arrows pointing up to the notional white spaces. If the company has allowed silos to be erected, there will be white spaces to be bridged between these as well as between functions, departments and groups.

Sometimes communication, decision-making or both is confined to silos, too. It is not unknown for the equivalent of George Cheddar, say, to insist that all important matters go through his office instead of direct to marketing or production. Intranets and internal email have undermined such centralist edicts in most computerized organizations. The result is that 'doers' can speak direct to other doers. (Whether they do so too much is another question, not for consideration in this book.)

The next version of the Cheese Inc. chart (figure 6.7) shows the path of two processes that span the organization. The order process goes from left to right and the supply process goes the other way.

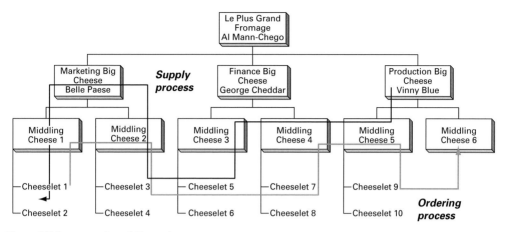

Figure 6.7 A process view of Cheese Inc.

If this were a diagram of real processes, we would want to know why they both apparently meander so much. They take curious-looking routes, and to unexpected organizational levels, to achieve their purpose.

It could be that these wobbly lines are justifiable. Outline diagrams can disguise as much as they make clear. We would want to know more about data (sources, outputs and sharing), communications flows, tasks, roles and timings before thinking about whether it is possible to improve these processes. We would also look at the interconnections between processes and whether these could be improved. The white space between processes is as important to manage as that between organizational entities. There is always white space.

Rummler and Brache emphasize this in their book:

All organization structures have white space. The mission is not to eliminate white space. The mission is to minimize the extent to which white space impedes processes and to manage the white space that must exist.

Figure 6.7 above also makes clear the difference between an organogram and a process view of the organization. These wiggles might simply result from trying to overlay process routes on a map of an organization's formal power structure.

A famous example of how maps of the same territory can differ is the London Underground railway map, which first appeared in 1933. Based on an electrical circuit diagram, it distils out the process aspects of a 'Tube' journey. Every station appears on the map, which uses colour coding to show which service goes to which stations. Where more than one service goes to a station, the differently coloured lines converge or cross. The result shows the least information a traveller needs, in a stylized and simple form. Transport authorities all over the world have followed its example.

What was revolutionary about the Tube map is that it does not show distances between stations to scale. The directions shown on the map are distorted, too, not

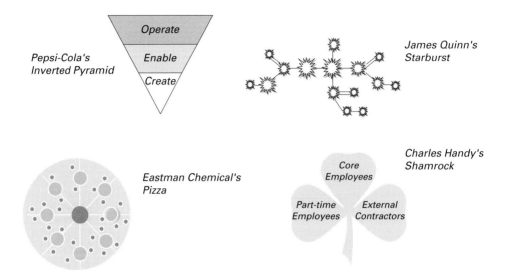

Figure 6.8 Some alternative organizational forms

being related to stations' compass bearings from one other.[12] The result looks strange when compared to a geographically true map but it portrays a kind of reality that is just as valid.[13] In the same way, the Cheese, Inc. process flow diagrams are as true and as valid as its organizational structure chart. Some might say they are more so.

The traditional organogram also exhibits what might be called the PowerPoint effect. Because charting software typically offers only this kind of hierarchical diagram, people tend to draw that kind. It reinforces existing stereotypes, which can be visual as well as verbal.

There are many other organizational forms available. Figure 6.8 shows a few possibilities that have been tried or mooted in the recent past. A process view of any organization adopting these forms would no doubt be as serpentine as that for Cheese, Inc.

Groups and teams

Another dimension to organizational life omitted from most organization charts is membership of workgroups, especially online groups. The workgroup has existing since pre-industrial times. It was the Stone Age hunting party, it is the military

---

12 All lines run horizontally, vertically or diagonally (at 45 degrees) – or consist of combinations of these.

13 Transport for London, which operates the London Underground system, has different versions of this map on its Web site. The site displays the 1993 original and the current version. It also offers what it calls the 'real Underground map', demonstrating how people are conditioned by looking at 'ordinary' maps. See http://tube.tfl.gov.uk/default.asp.

Table 6.3 Workgroups and departments

Element	Workgroup	Department or function
Structure	Often informal	Formal
How created	Formally and informally	Formally
Entry method	Sometimes formal and, if so, usually by appointment. If informal, casual and by common consent	Formal, bureaucratic and long-winded
Membership	Multiple (a person can be in several)	Single (a person is usually in one)
Tenure of membership	Highly variable, from hours to years	Usually months at least
Boundary	Porous; many overlaps	Sealed; little or no overlap
Lifespan	Days to years	Years
Location	Single, multiple or virtual (online)	Mostly single; sometimes multiple
Activities	Work (formally defined or otherwise); hobbies (e.g. company-sanctioned social club); subversion (see below)	Formally defined work
Span of attention	From a single group of desks to organization-wide and beyond.	The department or part of it

platoon, it is the football team. It is large enough to spread the load adequately, whatever that load might be, but small enough for everyone in it to know each other. From that knowledge arises trust and mutual confidence.

Workgroups are not the same as departments or functions, as table 6.3 illustrates.

Functions and department are features of the formal organization. The workgroup, conversely, is often but not always an expression of the informal organization. For instance, in figure 6.9 of some of the workgroups in Cheese, Inc., we can see a mixture of formality and informality.

The management board is formally appointed and permanent, as in most organizations. All the top managers, including the managing director (chief executive officer or CEO), are permanent, *ex officio* members of the board. Typically they will work alongside the company chairman as well as non-executive directors from outside the organization.

Also formally set up was the group to evaluate 'Cutting Edge', a range of kitchen utensils the company will be launching. Its members are there partly through position, partly through expertise and, possibly, partly through tokenism, representing some social sector ('demographic') or other. This workgroup is most likely to be semi-permanent, with a life measured in months.

The third collection, the self-styled 'Curdish Rebels', is a ginger group, dissatisfied with the company's direction and management. They are working on a buy-out

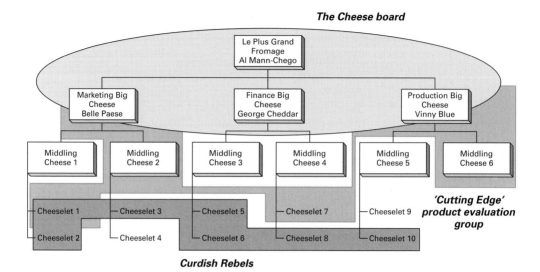

Figure 6.9 Workgroups in Cheese, Inc.

plan and are in contact with possible financiers. Being mostly young and technically aware, they communicate by unrecordable means – word of mouth (including personal cellular telephones), instant messaging and off-site, after-hours meetings.

There are two other workgroups not shown in figure 6.9. These are the participants in the ordering and supply processes. They might not see themselves as colleagues (another aspect we would look into) but they are bound together by the processes.

Helping process members to see themselves as part of a group is an important part of BPM. This might also apply to individuals and groups in customer organizations, suppliers and trading partners. Electronic commerce creates 'virtual teams', whose members physically meet seldom or not at all. (See the next section for more on this.)

BPM candidates at Cheese, Inc.

Which of the groups listed above should Cheese, Inc. look at if it were embarking on BPM? The answer we would give is that buying and selling take precedence, being essential to the effective daily running of the business.

On the evidence so far, the product evaluation group comes next. Its members' primary needs will be for good communications and access to shareable data. After that, they will probably want some process support, such as in distributing versions of product specifications. Modern process management products allow short-lived processes to be automated quickly and at little expense.

The board comes a long way after these. Although its members engage in some routine activities at meetings, their processing needs are slight and they will probably not want to use computers in meetings. One for the back burner.

And the rebels? If there is a revolution – and you survive it – efficient processes will be near the top of their wish list. They and their backers will want a cost-effective and adaptable organization so they can recoup their investment quickly. Until then, it is probably unwise to try to set up an underground BPM programme, even if you are one of the rebels. Besides, the auditable nature of computer-managed processes militates against the secrecy these people presently need.

The virtual organization

Silo-based thinking is unsuited to the fluid realities of tomorrow's organizations, and many of today's. Location has also become less significant, in response to the trends towards globalization, round-the-clock trading, mobile workforces, variable working hours and inter-company working.

Not every organization wants or is able to adopt such working patterns, but many do. There are three separate elements to this trend:
- the virtual workplace
- the virtual team
- the virtual organization.

The *virtual workplace* is what many nomadic employees are using at the moment. Through choice, need or compulsion, they have abandoned the permanent desk. They are instead using a combination of portable computer, fax machine, cellphone and temporary locations to perform the same or similar duties as before. Many are self-employed but an increasing number are employees. They are adopting this working style mainly to save accommodation costs for their employers.

Members of a *virtual team* work in fixed or virtual workplaces. The virtuality comes from their not sitting together, because of distance or time (different shifts in the same building, for example). These teams are the main beneficiaries of collaborative software.

The *virtual organization* is not ousting the permanent organization but is complementing it. It consists of permanent and temporary employees and permanent and temporary teams or units. Project organizations are typical. These usually consist of employees and units from the leading company, with companies, groups and workers from elsewhere collaborating with it and each other.

A variant is the extended but synchronized supply chain that is the goal of many manufacturing organizations. Other examples are the Internet, the *keiretsu* ('leaderless group') combines found in Japan and the international VISA card operation that Dee Hock helped establish.

Some industry pundits get intoxicated with these notions, and propose them as a practicable solution for almost everyone. Their enthusiasm needs tempering with reality. There are vast areas of economic activity, particularly in the extractive and manufacturing industries, that cannot go 'virtual'.[14]

No matter how far-flung the makers of its component might be, for instance, there comes a point in the life of a washing machine, a motor car or a house when all its parts have to meet, to assume a new and fixed physical reality. They are of no use to the customer unless and until they do. The same goes for clothing, furniture and food. These are tasks that cannot be 'virtualized'.

There are also jobs where the requirement to attend is inescapable. Doctors, police officers, nannies, shop assistants, receptionists, bank tellers and so on can all, to varying degrees, make some use of electronic communications but the essence of their work lies in direct personal contact. Nevertheless, the greater use of virtual working arrangements is real and has the potential to yield benefits if adopted imaginatively and sensitively.

To our mind, a more useful suggestion is that a virtual organization should not be seen as a distinct structural form. Instead it should be thought of as a characteristic of the underlying form, be it functional, matrix, inverted pyramid or whatever the strategic choice is.

Process-based structure

To whatever degree it is adopted, the virtual organization is a communications-based organization. Interaction and reporting become less dependent on physical location than was traditionally the case. Direct supervision and management are almost impossible.

There is nothing new in this. The Roman Empire, the Roman Catholic Church and the British Raj are all historical examples of successfully managing at a distance. The difference is that they measured time in weeks and months. Modern organizations operate at highly compressed, and diminishing, timescales. Management reports consisting of ship-borne letters travelling thousands of miles are not a luxury they can afford. Even the Vatican is on the Internet these days.

Where novelty does arise is in the perception of an organization as a conduit for work processes. In this view, processes run through it like wires through a main telephone cable. Each carries its own traffic and pops into view only where it needs to connect to something in the outside world. Good management consists in ensuring there are no kinks in the wire, the connections are sound and signals travel down it at maximum speed. This normally means making the wire as short as possible.

---

14 As we said earlier, all organizations are 'virtual' to a large extent, having no physical existence. The word is sometimes used without thought, much as other vogue expressions like 'agile' and 'lean' are. (This results in such silly expressions as 'agile employees' and 'lean contractors'.)

This analogy breaks down when you consider the subject of insulation. Telephone engineers work hard to prevent signals leaking from one wire to another, in what they call crosstalk. Organizational processes that were similarly insulated from one another would merely replace vertical chimneys with horizontal flues. Without crosstalk between processes, further improvement and better all-round information become impossible. (Please note that we are not here arguing for complete openness of information flow between functions. Such an idealized state is humanly improbable and politically suicidal.)

Achieving such maximal flow entails other changes. Here is how Michael Hammer and James Champy describe the situation in *Reengineering the Corporation*. The italics are theirs:

In the long run, reengineered processes can only thrive in a *process-centered organization*: one in which process owners are not transient project managers, but key executives charged with assuring the long-term health of the processes; where measurement systems focus not on functional performance, but on total process performance; where people's compensation is linked to how well their processes perform; and where all people understand the company's processes, and how their individualized work contributes to realizing the company's goals.

It is difficult to argue with any of this; such hope and wholesomeness is hard to resist. Yet there is the feeling that there is a baby somewhere being horizontally ejected with the vertical bath water. What happens to product responsibilities, strategy, team building, accounting and other activities that are not part of work processes? Even if these can be adapted to horizontal modes, how costly and painful will the change be?

What happens to other processes, such as management, marketing and learning? And where do middle managers go? They are still useful and often necessary to maintain continuity of knowledge and culture, for example. Are all types of organization and every part of any organization suitable for swivelling through 90 degrees?

Adopting absolute positions on questions like these is a demonstrably successful way for consultants to sell books and launch careers. For practising managers it is neither wise nor, often, possible. They must test such radical ideas against their own experience and the messiness of real life. Instead of making dualistic choices – it is either this or else it must be that – long-lived companies position themselves somewhere between extremes. Moving towards a process-based structure is, in our view, an option that any organization should explore. It may even decide to evolve towards such a form. Many do.

Becoming a 'horizontal' organization is not, however, always necessary in becoming a process-based organization. So long as functions, departments, job roles, locations and other 'vertical' elements do not impede the flow of processes, it is safer, quicker and cheaper to keep them as they are. If, on the other hand, they are removed solely because of some doctrine, it is fair to ask who is running the organization – its managers or an unknown and unaccountable 'academic scribbler'?

Case Study 5 Morse Group Ltd

Originally a personal computer retailer, the British Morse Group has grown into a Europe-wide supplier of systems and integration services. Some of its internal processes had not kept pace with this growth and were overdue for improvement. Its sales support system was the first candidate.

Since introducing new systems is what Morse does for a living, it decided to run most of the improvement project itself. Having selected a supplier, the company's internal systems staff installed the process management software in parallel with some new corporate software from SAPAG. The two programs reinforce each other.

The project team involved managers and users throughout; indeed, the project arose from a suggestion by users. The team engaged in collaborative and iterative development, in which users were asked to try each new evolution of the system as it emerged. As with some of the other cases, the company appointed process managers to take ownership of specific aspects.

Results have been clearly satisfactory. The processes concerned are now well managed, consistent, adaptable and expansible. Compliance with ISO requirements is eased, the system acting as 'a living ISO9000 manual'. Sales staff no longer have to resort to faxes and repeated telephone calls to resolve customer queries but deal with them online. Customers are noticing an improvement and have said so. Morse plans to apply the software to several other processes.

Industry/ Sector	Computing services	Location(s)	United Kingdom HQ
Annual turnover/ income	£351 million	Number of employees	1,100
Type of system	Workflow	Supplier and product	Metastorm e-Work
Number of users	About 300	Time to complete	5 months
Business objectives	• Standardize internal procedures • Improve employee productivity • Increase customer responsiveness		
Quantitative results	Not quantifiable – the business environment was too unstable to allow meaningful before and after comparisons		
Qualitative results	• No lost purchase orders • Fewer customer queries on purchase order status • Has been the IT project best received by users • Fewer bottlenecks in the system • Clear lines of ownership across the supply chain • Unprecedented visibility of delivery information for warehouse staff • Improved customer perception and satisfaction		

Business background

Morse Group Ltd began in 1983 as a retailer of PCs. It has grown and progressed to become a leading system integration and computer services company. The company provides a one-stop service to corporate clients, installing and managing complex applications and systems for them. Morse's objective is to make it easier for its clients to achieve their business goals by using computers.

Its clients include multinational organizations operating in Britain, France, Germany, Ireland and Spain. They come from a wide range of industries, including wholesale and retail financial services, telecoms, media, entertainment, leisure and the public sector. Nearly half of Morse's clients have been with it for over five years.

The company is ISO 9001:2000 certified by the British Standards Institute (BSI). The scope of the certification is given as: 'The supply of IT consultancy and the delivery of configuration, installation and servicing of electronic computer systems equipment, including sub-systems.' The company places great importance on rigorous internal audit processes. It sees these as fundamental in continually improving customer satisfaction, internal efficiency and employee involvement.

Morse identified three areas of its business that it needed to improve to maintain service levels. These were order management, customer quotations and pre-sales resourcing. The existing processes for all three were time-consuming, people-intensive and ripe for efficiency improvements.

Order management and tracking, for example, was mainly manual. Employees keyed customer order details into the existing SAP system. They later had to check the SAP system and with the staff involved to see if transactions had been completed. It was difficult to oversee the progress of orders. Paperwork also got lost, causing delays and customer dissatisfaction.

To streamline these processes would need a system that could handle complex, diverse procedures and that could evolve as they did. It also needed to be readily understood and embraced by Morse's staff. Equally important was the ability to optimize the existing IT infrastructure, particularly its SAP system.

Morse thus had three aims:

- to replace its existing systems
- to provide abilities not available in its existing systems ('filling the gaps')
- to provide abilities available in its existing systems but whose introduction could not be justified in cost terms.

After a two-month evaluation of several offerings, Morse chose Metastorm for its affordability, flexibility and ease of integration.

System description

The first target of the workflow system was the order management system. This affects all areas of Morse, such as bench engineering, warehousing, buying and credit control. (Bench engineering is the building and configuring of equipment.) Previously, staff entered sales details on to paper with the purchase order attached. This then passed through the different departments until an order was completed.

Now, when a purchase order arrives, staff first scan it into the Metastorm system. Everyone involved in order fulfilment can then see that image and details of the progress the order has made. They can consult this information together with SAP reference data. This new and up-to-date information lets them deal with queries directly and immediately. Other details available online include order history and the status of workloads across all staff and departments (figure C5.1).

The workflow system enables Morse to deliver products to customers faster and ensure more efficient payment. Not a single purchase order has been lost since introducing the system.

Another important process to be automated was managing price quotations for customers. These are created in SAP. Previously, staff would fax a quotation to the relevant consultant, who could be anywhere in the UK. The consultant would check its accuracy and technical configuration. Resolving queries entailed telephone calls and further faxing.

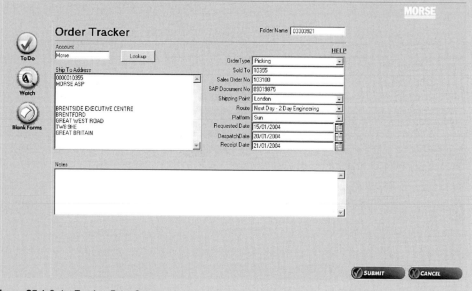

Figure C5.1 Order Tracker Entry Screen

Using the workflow software, Morse sales staff can now send quotations electronically to consultants. They deal with queries online. If staff do not hear back from a consultant within a set time, escalations built into the system alert them to act. They can then ask consultants from other offices to verify the quotation. This gives a smooth transition of work, compliance with customer service-level agreements (SLAs) and adherence to ISO demands.

In the past, Morse employees needing pre-sales resources needed to check if a pre-sales consultant was available and complete a form to book his time. Finding an available consultant often took several telephone calls. Booking now takes place online. This frees them to spend time on the tasks they were employed to do.

The system serves about 300 users, in Morse's British and German offices. It runs on a single server, under Microsoft Windows 2000, and links to:

- Crystal Reports for management reporting
- a custom-made document scanning module
- SAP.

Implementation experience

Where and when did the project or system originate?	A user department first saw the need for a work tracking system. Rather than develop a bespoke solution for this one process, the IT department proposed the use of a general workflow management system that could improve business processes across the company.
How long did implementation take?	5 months from project initiation to implementation. The project was dependent on the progress of a parallel project of installing SAP software
Who did the implementing (own staff, contractors, consultants, etc.)?	About 10 days of external consultancy were used during the development of one procedure. This helped transfer knowledge to Morse staff. All other development was undertaken in-house.
How much bespoke development was there?	It was all bespoke
Were there any special infrastructure needs?	None, other than some new scanners.
What were the most significant implementation issues and how were they dealt with?	Morse worked hard at preventing these. It drew up a project team drawn from all areas of the business likely to be impacted. 80 per cent of the work was in defining and agreeing the workflow, related business processes, responsibilities, authorization levels, etc. Only 20 per cent of the effort involved development, testing and roll-out work.
Who is responsible for the system overall?	• Each procedure has an owner who is responsible for ensuring that it remains current and relevant to the needs of the business.

	• The system is almost entirely self-managing; little human intervention is required for administration.
	• The internal systems department is responsible for the development and updating of procedures, setting up new users, etc.
How was and is training handled?	The user members of the project team became 'power users' and trained the other users.
What was and is done to encourage use?	In the implementation phase, members of the project team were encouraged to demonstrate early prototypes of the procedures to their colleagues. Feedback and ideas were gathered, responded to and, in many cases, actioned. It therefore became their system and in their interests to ensure that it worked.
	Since implementation, the benefits derived by the users drive the use of the system.
What lessons were learned?	• Establish and agree the workflow, related business processes, responsibilities and authorization levels before starting development.
	• Use prototypes of the system.

Successful implementation of the new system required business groups to work closely with the IT department to define the process. Cross-functional projects like this need to have their ownership clearly defined.

The Metastorm software was quickly installed, with minimal disruption to employees. Because this could be done in stages, Morse could keep track of the project's progress before extending the system.

Benefits and user reaction

What has been the reaction of managers and staff?	The system has been well received and accepted, as a result of the approach described above. Those who were its biggest critics initially are now its greatest evangelists.
What has been the reaction of customers or trading partners?	The system itself remains largely transparent to customers or trading partners. They see only the benefits that derive from using it.
What has been the overall cost of the system?	£60,000 in Metastorm software licences, £10,000 for external consultancy and £10,000 for computer hardware.
What have been the main process benefits?	• Clearly defined processes that have been agreed across the business.
	• Better visibility across the business.
	• Measurable SLAs.
What have been the main effects on operating style and methods?	Work is always driven forward. Responsibilities are clear.

The new system has increased work throughput, taken paperwork out of the business and enabled an up-to-date view of activity, regardless of time or location. There is now no paper to route, store or lose.

A major benefit, although hard to quantify, is the improved customer service. Faster customer response times have meant an overall improvement in customers' opinion of Morse.

Workflow automation has given Morse consistent, visible, scalable and auditable business processes across the organization. The Metastorm software offered a practicable and low-cost route to improving business processes that could be introduced by stages.

In summary, adopting workflow automation has given Morse:

- explicit definition of business processes
- increased consistency and fewer process errors
- adherence to ISO service levels, the system acting as 'a living ISO 9000 manual'
- better management of service level agreements
- data validation at point of entry, leading to less rework
- enforced authorization
- visibility of the current status of each transaction
- an automatic audit trail
- integration of existing systems, such as SAP, adding to their value to the company
- easily changed processes, to adjust to business changes
- the flexibility to work remotely anywhere in the world, with the same information and control as back at the office.

The future

The company plans to extend the use of the Metastorm software to other processes, such as:

- non-stock buying
- stock returns
- customer feedback tracking
- incident reporting.
- invoice approval
- installation request, scheduling and management
- HR management actions on starters and leavers.

Morse is also working on a second version of the sales support system.

7 The business return on BPM

Introduction

BPM can provide organizations with rich benefits, which stem from various sources:

- the greater control that managers have over existing processes
- the ease with which these can be varied or new processes introduced
- the improved coordination of activities within and between organizations
- the simplicity with which new computer systems and services can be integrated with existing arrangements
- the new or newly applied business models that BPM makes feasible
- the reduced costs associated with all these.

These various gains do not need extolling here, and many organizations are already experiencing them. Rather, in this chapter we examine the underlying principles needed to get those benefits. We look at how organizations can analyse, justify, direct and monitor their investment in their process improvement programmes. Also, we ask how relevant the return on investment (ROI) approach is to assessing such investments.

Early in the chapter, we discuss an often-overlooked part of making a business case – the motives behind investing. Our view is that these should drive your choice of computer system, design methods and target processes. They should also guide your choice of justification method, which is not always financial. The diagnostic approach we describe can apply to any kind of investment, not just in process management or even in computers.

As in other places in this book, we use the word 'business' irrespective of whether an organization is a profit-making entity.

The contribution of computer investments

Delusions and paradox

People grow to see the world through the distorting lens of their job specialization or discipline. IS specialists are especially susceptible to this occupational squint. It shows itself in two erroneous beliefs, which some people seem to treat as axioms or 'givens'. The first delusion is that the performance of computer systems is the major determinant of business performance. The second, stemming from this, is that better information systems always lead to improvements in business performance.

You may feel we are exaggerating, that nobody is that naïve. We would point you to the wide currency given to the expression, 'productivity paradox'.[1] This has given rise to countless white papers, consultants' reports and academic theses.

The supposed surprise is that putting computer investment into a process does not reliably lead better business performance. Some people believe (or affect to believe) that this flies in the face of common sense, hence the so-called 'paradox'.

Now, according to the *Oxford English Dictionary*, a paradox is 'A statement or tenet contrary to received opinion or belief; often with the implication that it is marvellous or incredible'. There is nothing marvellous or incredible about the non-arrival of the planned benefits of any computer systems change. It is a disappointment, naturally, but also an everyday fact of life – just ask any CIO or most British government departments.

The potential obstacles to gaining system benefits are numberless. Poor training, a change in workplace layout, an organizational reshuffle or an unexpected new focus of managers' attention can all derail expected improvements. These are problematical, to be sure, but not paradoxical.

Only the greenest IT manager or consultant can be unaware of success' dependence on circumstance. Perhaps we are here in the land of official beliefs versus unofficial, or possibly it is professional pride at stake. Whatever the cause, such denial of reality does not help with an honest examination of the causes and consequences of investing in process management. As the philosopher and theologian, Peter Abelard (1079–1142), put it: 'Constant questioning is the first key to wisdom. For through doubt we are led to enquiry, and by enquiry we discern the truth.'

That questioning should also apply to a third belief, that better business performance leads direct to greater business success. This is the grandest delusion, exhibited daily in press releases, magazine articles and conference presentations. Business success, more even than project success, is uncertain. It depends on senior managers' skill, fashion, stock market fluctuations, national and international politics

---

1 Google reported nearly 12,000 occurrences of it on the Web alone.

and many other factors, not the least of which is luck. Good systems and processes help an organization exploit favourable conditions and weather unfavourable ones, but they are not determinants of success.

Paul Strassmann has studied and written about the reality of computing benefits for many years. In his book, *Information Payoff*, he examined information technology investments in the light of what he termed 'management added value'. In other words, he looked at computers in terms of how they help, or hinder, management performance.

Strassmann came to the conclusion that IT is an amplifier of management performance. In other words, a well-managed company will gain strategic advantage from its computing investments. Conversely, a poorly managed company will just make matters worse for itself by spending money on computers. Moreover, the more such companies spend, the worse matters get. You can probably think of examples of both results from your own experience.

St. Thomas [Aquinas] had a sound mind, and it was he who really taught us to distinguish without separating.

So wrote the philosopher Jacques Maritain, in *The Peasant of the Garonne*. Distinguishing without separating is at the heart of this book. You cannot safely divorce in a practical sense any single element of the organizational organism from the other elements. These components can, to an extent, be examined as if they were unconnected with one another but even there such reductionism is risky. This is as true in investment analysis as in any other realm.

Why invest?

In business, there are three primary motives for making any change:
- to do much as you do at present, but *cheaper*
- to do much as you do at present, but *better*
- to do *new* things.

These apply separately and in combination.

In the first of these, *cheaper* organizations will typically want to avoid or reduce costs such as labour, materials or bought-in services.

The *better* motive can be divided into two. The first deals with internal operations, and introduces greater control, better decision making or fewer errors. This is *better internally* and is to do with improving capability. The other looks to improve external working in areas such as customer support, delivery times or product quality. This is *better externally*, to do with improving performance.

Then there is innovation, such as in operating methods, organizational structures or markets served. This, obviously, is *new*.

Table 7.1 Specimen investment actions and expectations

Cheaper	Better internally	Better externally	New	Compliance	Infrastructure	Image
Fewer/cheaper employees	Faster processing	Better quality of output	New products or services	Quality management compliance (e.g. ISO9000)	New/replacement backbone systems	Corporate image
Increased output for given cost	Faster/better decisions	Faster response to customers	New customer relationships	Group EDI system	New/replacement networks	Attractiveness to potential staff
Less/cheaper material	Fewer delays	Shorter time to market	New information sources/outlets	Environmental-impact compliance	New/replacement software development environment	Demonstrating capabilities
Reduced stock-holdings	Fewer errors	Reduced exposure to risk	New organizational structure	Building regulation compliance		Attractiveness to potential customers
Faster start-up of process or project	Better control	Better information to, from and about customers	New working arrangements or methods	Data protection compliance		To boost confidence of staff/customers
Fewer/cheaper bought-in services	Better/faster management information	Better service	New markets	Audit compliance		To worry competitors
Extending life of systems or equipment	Better resource allocation	Better support of customers	New external relationships	Security control system or rating		To impress potential trading partners
Less scrap or waste	Improved staff/systems availability	Faster delivery times	New functions performed	Health and safety regulation compliance		
Faster invoicing	Smoother work-flows	Longer service hours	New distribution methods			
Better credit control	Better integration with existing processes/data	Closer links with trading partners	New trading arrangements			

For BPM, typical objectives might thus be:

- cutting the costs of information provision and searching, of customer interaction and of transactions (cheaper)
- speeding the flow of internal information, improving collaboration among functions (better internally)
- improving speed, reach, scalability, interactivity and contact with customers and trading partners (better externally)
- gaining new customers, entering new relationships and markets, finding new routes to market and creating new products (new).

Although the cheaper–better–new trio dominates, other motives occasionally arise. Organizations may wish to:

- *comply* with mandatory requirements, such as audit or legal standards (for example, ISO 9000 or pharmaceutical records retention)
- provide an *infrastructure* for other activities or investments, such as integrating front and back office systems, or
- improve their *image*, typically in pursuit of branding.

Table 7.1 shows some examples of the kind of actions and expectations you would expect to find in each of those seven categories.

These seven motives are not mutually exclusive and they often apply in combination. They are also universal, being equally applicable to corporate take-overs, buying a fleet of fork-lift trucks or, indeed, changing corporate processes.

In addition, the motives vary around the organization and change with time. Figure 7.1 shows their relationship with one another.

For ease, we refer to the outer set – cheaper–better–new – as the first-tier motives and the others as second-tier.

Bear in mind that this model refers only to primary motives. You could argue, for instance, that you were making a particular infrastructure investment in the hope of improved cost or quality. That would be reasonable, and is common, but the latter two motives are secondary. They would result from the infrastructure investment but not be the immediate target of it.

The investment cycle

One reason for showing these relationships diagrammatically is that it lets you show priorities more clearly than in a list. We happened to mention 'cheaper' first and 'image' last but the sequence means little. These seven motives are equivalent to one another. One is not always better or more important than the other. Their significance depends solely on their appropriateness to the particular time and circumstance.

Having said that, there is often an order in which the outer four motives arise. Most organizations' objectives for using any tool, structure or method tend to evolve along a similar path.

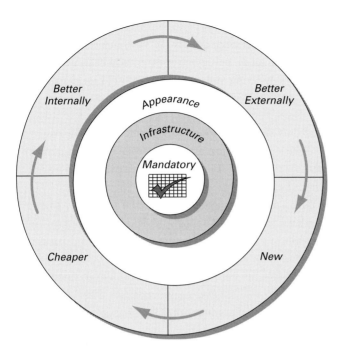

Figure 7.1 Virtuous investment circle

A common first objective in adopting new computing systems is cost saving. Cost reduction and cost avoidance are favourite reasons for embarking on almost any business process intervention. You have only to look around your own organization to realize that existing businesses nearly always introduce changes with this uppermost in mind.

Later, having made the savings (or, far too often, having discovered that they were illusory), businesses typically look at ways of improving their internal operations. In other words, they wish to 'run a tighter ship', 'create a more streamlined organization' or some other figure of speech. In short, they wish to be better internally.

This sometimes comes about accidentally. Organizations introduce a system or make a change on cost grounds. They then often find that the processes concerned are running smoother or faster or there is some other windfall benefit. This phenomenon is known as the 'emergent properties' or 'second-order effects' of the system. Having discovered that possibility, the organization will typically then set out to exploit and extend its accidental gains.

These two sets of motives – cheaper and better internally – are both inward-looking and, often, backwards-looking. Until the days of the World Wide Web, they tended to predominate in IT investments. It was usually only later that a business looked outwards in its thinking about its computer systems.

At that stage, a typical objective would be to improve dealings with customers and trading partners. Issues of timeliness, quality, service and responsiveness begin to assert themselves. In other words, being better externally becomes important.

After that, there comes applying systems to something new. New products, new working arrangements and innovative ways of dealing with customers are all typical of this phase.

Innovative applications are often those that yield greatest competitive advantage. Despite this, until recently, innovation was hardly ever the starting motive for any computer investment. It is still not common in many businesses. (See Diagnosing introverted investment below.)

Arriving at innovation does not signal an end to the sequence. As circumstances change, and as yesterday's pioneering application becomes today's routine, the cycle starts again. After a while, a business will look at its once ground-breaking systems. It will then typically decide that it is time it started cutting some of the costs of these.

A business first consolidates its innovations and then looks to improve on them. It works through the sequence over and over again, endlessly. The 'virtuous spiral', as it now shows itself to be, is a model of organizational learning. It is illustrated in figure 7.2.

Timeless motives

We have not yet discussed the second-level motives, those at the centre of the 'virtuous circle'. These are the mandatory, infrastructure and image motives. In contrast to those in the outer ring, they can arise any time in the investment cycle. (Their position relative to one another in the diagram is irrelevant, therefore; it is just the way we have drawn it.)

The innermost circle consists of mandatory or compulsory investments. These are imposed upon the organization, usually by a government, regulatory body or trading partner. A common example is compliance with national or international standards, such as for security or quality.

Government legislation often imposes new compliance needs. Examples in America include the 1996 Health Insurance Portability and Accountability Act (HIPAA) and the 2002 Sarbanes–Oxley Act on corporate financial reporting. In Europe, the Basel II rules for bank capital will impose data management demands that must be met by the end of 2006.

Compulsion also often arises among supply chain partners. For instance, EDI users must adopt certain standards if supplying information through that medium. If they are not the dominant partner in the relationship, they usually have no choice in the matter. They either make the investment or give up the chance to trade with the larger organization. This is starting to reappear as a motive with the rise of Web services. (Compliance with internal, self-set standards does not count here. It lacks the element of external compulsion that makes it mandatory.)

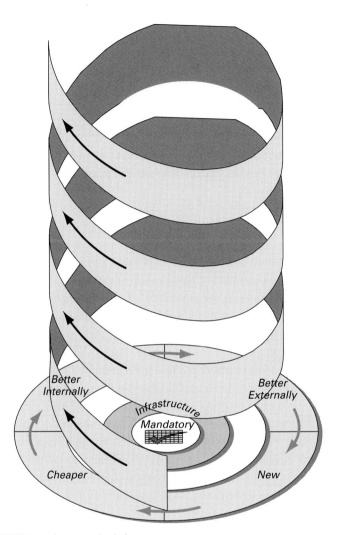

Figure 7.2 Virtuous investment spiral

An infrastructure investment, in the next ring out, might not be recognized as such at first. Installing an electronic mail system, for instance, may have been done with one or some of the other motives in mind. Only later may it become clear that this can now act as a messaging service. Other applications, such as a CRM system, might communicate over this. Further investment in the email system, therefore, would be classified as infrastructural. It is the motive that determines the type of investment, not the technology invested in.

Immediately underlying the outer four motives is appearance. This applies whenever an organization invests to impress, whether it be its customers, trading partners or employees. Indirect marketing, such as corporate advertising, is a typical image investment, as is anything concerned with branding. Image enhancement is also the reason behind companies' demonstrations of capability, such as computer suppliers' conspicuous use of their own products. You will not see BMW's executives driving a Ford while on company business for the same reason, even though it would probably be cheaper. The difference in vehicle cost is an investment in image.

Personal image improvement on behalf of a department or an individual manager is another commonly encountered motive behind change projects. It is not often owned up to, though, and is usually officially disapproved of.

Using the 'virtuous circle'

The following section shows how you can apply the 'virtuous circle' model to statements of benefit, both promised or expected. You can also do so to results, as an after-the-fact check. Translating these various statements into the terms of the 'virtuous circle' shows you which of them has a solid foundation of thinking.

We examine here the system investment motives for three organizations – iJet Travel Intelligence, BT Exact and Availity. These companies are from different industries, with different business models and use different BPM products.

The information we show is all in the public domain. We have purposely chosen published material rather than projects we have been involved in. This is to show that you can apply the framework to any set of benefit statements even without first-hand knowledge of the background. Working 'blind' like this is, after all, what senior managers and directors often have to do when they are asked to sign off company projects.

Imagine how much more penetrating your analysis would be if applied to an organization you know, perhaps your own.

Snapshot 7.1 iJet Travel Intelligence

iJet Travel Intelligence is a privately held company, based in the USA. It provides travel intelligence and risk management services. These go to various industries, especially travel, insurance, mining and oil extraction. iJet also works into US government departments.

The company began trading in 1999 and employs about 100 people. Its computer systems use products from Oracle, BEA and Sun, as well as bespoke software for searching, pattern recognition and speech-to-text conversion. Its BPM software is Fujitsu-iFlow.

The essence of iJet's service is the round-the-clock supply of information on potential and actual travel problems. Aided by sophisticated software, multilingual teams of analysts sift this from a continuous inflow of data from world-wide sources. Experts on varied subjects – geographical, political, commercial and so on – review and edit this intelligence before it goes to customers. It reaches them mainly through email and the World Wide Web. Individual travellers can also get it by satellite and cellular telephone.

As you would imagine, speed of response inwards and outwards is of primary importance to iJet and its customers. Stale information is of no value to it or them. The company also prizes the ability to adapt quickly to changing business demands and opportunities. It has done so several times, sometimes in just hours.

iJet's chief technology officer (CTO) has defined this company's expectations of its systems, as:

- automating complex business processes end-to-end
- integrating applications to carry out process steps
- opening up processes for collaboration among staff, customers and partners
- providing a framework for streamlining workflows
- integrating easily with the existing IT infrastructure
- giving the necessary ease of transformation of its structure and processes
- ensuring that all editorial and approval guidelines are followed
- ensuring that the travel intelligence is rapidly in the hands of customers.[2]

Translating these into 'virtuous circle' motives produces this result:

- 'automating complex business processes . . .'. This is a systems requirement, not a business objective. Automation does not necessarily produce a business benefit. If done badly, it could even be damaging.
- 'integrating applications . . .'. As with the previous statement, we can assume this to be better – internally, externally or both – but we cannot tell. This also is a systems requirement.
- 'opening up processes for collaboration . . .'. This appears to be better (internally and externally) and possibly new. It, too, needs refining.
- 'providing a framework . . .' is infrastructure.
- 'integrating easily with the existing IT infrastructure'. Another systems requirement. This could possibly be better internally and, at a guess, cheaper.
- 'giving the necessary ease of transformation . . .'. This is better internally and probably new.
- 'ensuring that all editorial and approval guidelines are followed' is a quality issue and is therefore classed as better externally.
- 'ensuring that the travel intelligence is rapidly in the hands of customers' is about service levels and is, thus, better externally.

---

2 From 'Jet-Fueled', a case study by Greg Meyer in the February 2003 issue of *Optimize* magazine (see http://www.optimizemag.com/issue/016/leadership.htm).

Some of our comments may seem harsh but we do not mean them as criticism of the company or its managers. iJet has made a success of its use of BPM software and of its computing resources overall. Indeed, it has won at least one major award for its process systems.

What we are instead showing here is how hard it can sometimes be to express systems needs in business terms. The material we have used comes mainly from a published interview with the company's CTO. He and his board-level colleagues will undoubtedly have done the necessary business testing of these systems criteria. The results of that test are not in the public domain.

Snapshot 7.2 BT Exact

BT Exact is the research, technology and IT operations business of Britain's largest telecoms company British Telecom. It is managing a BPM project in BT Wholesale, another BT division. This group provides network capacity ('bandwidth') to BT and other telecoms providers. It employs about 5,000 people, based in regions, who work with mobile and fixed-line operators. The group's business processes were complex and giving rise to a slow time to market, inflexibility and high cost. They were also unsuited to the demands of e-business.

The project uses products from IDS Scheer and Staffware. It was still under way while we were writing this book but had already achieved significant business benefits: The project's objectives are to:

- end the 'stovepipe' approach to systems development
- overcome the limitations of incremental development of systems
- agree national standards for records
- provide systems capability for geographical independence
- systems design to support radical organizational change options
- intelligent external inventory system to facilitate auto-planning.
  Some benefits have already been achieved, including:
- annual cost savings
- process standardization
- geographical independence
- 'multi-skilling' (of staff)
- improved time to market.
  These all translate to first-tier motives:
- cost savings equals cheaper
- process standardization equals better internally and, possibly, externally
- geographical independence is, we surmise, new
- multi-skilling is better internally and, possibly, new; cost saving is a probable secondary motive here

• improved time to market, is undoubtedly better externally.

Also, although this is not mentioned by BT Exact, the systems capabilities being introduced will form an infrastructure for future changes.

These statements are from a technical presentation by the project's director, who is a technical person. It would not take long with him, we feel, to get them all worded in business terms.

Snapshot 7.3 Availity

Availity, based in Florida, USA, provides a state-wide information exchange for health care organizations. The Availity Gateway is a secure Internet-based clearing house and portal. It links nearly 20,000 doctors, 6,000 practice offices and all 208 hospitals in Florida. Over 10,000 insurance claims pass through it every hour and it carries more than 20 million EDI transactions a year.

Federal legislation – the HIPAA of 1996 – is driving revolutionary changes in the American health care industry. The Act has two aims, the first of which is to reform the health care insurance market in the USA. The other aim is to simplify the processes of health care administration and reduce their cost. The Act imposes the use of standardized electronic ways of transferring administrative and financial data. This became compulsory in October 2004. It is this aspect that applies here.

Availity was set up five years after HIPAA was passed. Its founders have always known, therefore, that its systems and practices would need to comply with the law's demands. This, though, was just the price of entry to the game. Success would depend on going further.

One example is its systems integration programme. Availity is trying to link all its trading partner's practice management systems (PMS) with its own batch transaction and EDI systems. So far, it has done so with over forty PMS. It uses Vitria software for this and for its BPM work.

In its CTO's words, Availity's systems help it:
- facilitate the secure exchange of information between health care providers and health plans, regardless of the practice management system format
- decrease administrative complexity by providing a single point of contact for health care transaction processing
- reduce the average claim processing transaction costs for providers and payers by more than two-thirds
- enable HIPAA [US government] compliance for information security and secure electronic transactions for any provider that uses the Availity Gateway
- improve the patient experience by expediting procedure and referral authorizations, as well as claims submission and approval
- strengthen relationships between providers and health care plans by providing a single trusted intermediary

- shorten time-to-market by completing transactions five times faster using the Vitria solution.[3]

Applying 'virtuous circle' motives produces the following analysis:

- 'facilitate the secure exchange ...' is better externally and, for some trading partners, probably new.
- 'decrease administrative complexity ...'. This also is better externally.
- 'reduce the average claim processing transaction costs ...' is plainly cheaper.
- 'enable HIPAA compliance ...' is equally plainly mandatory.
- 'improve the patient experience ...'. This is better externally, again.
- 'strengthen relationships ...'. This, too.
- 'shorten time-to-market ...' As is this.

There are three things to note about this set. First is its clarity, second is its outward focus and third is its actuality. These are not promises or hopes but findings (or claims). They are framed in business terms and speak the customer's language.

It would be invidious to compare these statements directly with either of the other two. These came from software case studies, whereas those for Availity are from a business case study. All the same, we like the Availity CTO's ability to talk, and thus think, in business terms. It is a sadly rare gift among technologists, which is why the 'forcing mechanism' of the virtuous circle can be so helpful.

Extracting the message from the noise

You may think that we have been hard on one or two of the unwitting contributors of these cases. We would disagree. What we are looking at here is the way they have described their projects, not how they have conducted them.

Statements that are not easily or clearly expressible in terms of the seven motives demand caution on your part. They might be part of a set of promises, perhaps from a supplier or an internal provider. If so, you should question the person making the promises until he can define them in virtuous circle terms. If he cannot, or will not, then you should doubt the value of what you hear. Intentionally or otherwise, this person is purveying bovine ordure.

The same applies to any after-the-fact reports on benefits gained. Unless it is politically expedient to do otherwise, you should be similarly stringent about these.

The principle behind all this is simple – clear language stems from clear thinking, which leads to clear action. By demanding clearer language, you can induce that clear thinking or confirm its presence.

---

3 This list comes from a Vitria case study of the Availity installation, 'Availity Painlessly Delivers Collaborative Healthcare'.

Diagnosing introverted investment

It is usually revealing to get a group of executives to critically examine why their organization has made the IT investments it has. We have done this many times in closed workshops. Here is how you go about it:

- first, ask participants to nominate a major systems investments their organization has made
- get them to call out the reasons for making it
- list these on a whiteboard or flipchart, then
- explain the seven investment motives scheme.

Now you ask the group to call out which motive or motives apply to which reason they have suggested.

Having done this, you collectively 'score' the answers according to which motives apply. Allot one point to each mention of an applicable motive, whether it be on its own or in part of a group. Multiple motives are common.

In the BT Exact set, for example, the scores are:

- cheaper – 1 definite and 1 possible
- better internally – 2 definite
- better externally – 1 definite and 1 possible
- new – 1 definite and 1 possible
- infrastructure – 1 implied.

You can even, for greater effect, plot the responses on a 'virtuous circle' diagram (figure 7.3).

Like those for the other two case study organizations, the result for BT Exact shows a balanced approach, with plenty of right-hand as well as left-hand investing. This is no surprise. Any organization that is using BPM software well is, almost by definition, concerned with its speed of response and adaptability to change.

This is not the case in some kinds of organization. In pre-Web days, you could almost guarantee a different result for most medium-sized or large businesses. Their scores would fall mainly on the left-hand side of the virtuous circle (that is, better internally and cheaper). No more than 5–10 per cent of mentions would be on the right-hand side (better externally and new).

The remaining mentions would relate to the central circle (mandatory, infrastructure and appearance). These are, in a literal sense, even-handed motives, although you could argue that some 'image' projects are outward-looking.

Historically, decision-making about computer investment has been inward-looking and focused on local rather than global projects. This has had a distorting effect on the kind of project allowed to go forward. Anything on the right-hand side of the 'virtuous circle' has a hard time gaining approval in some organizations.

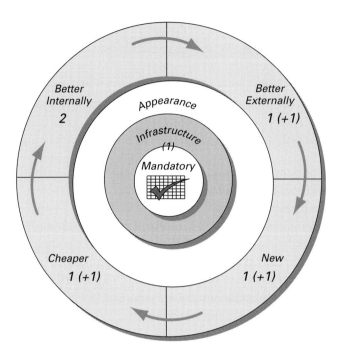

Figure 7.3 Sample workshop score

This, in turn, has had an adverse, but often unrecognized, effect on those organizations' ability to deal with external matters or respond to change. The current US obsession with shareholder value is having a similar effect.

Overall scores weighted to the left suggest introverted computing efforts and investments, commonly found in mature organizations. Managing and maintaining their computer systems is a complex matter on several fronts – technical, financial and political. It is therefore easy for the IT Director or CIO to be dragged into an overemphasis on these internal aspects. Managing them becomes a game in itself, to the detriment of broader business considerations.

An ideal ratio?

If there is this sort of imbalance in the results of your workshop group, news of it may come as a shock to some participants. Denial, then a demand for recalculation, will be frequent responses. The latter is easily accomplished but seldom changes the result significantly. Time and peer pressure usually cure the first.

Typically, the next response from participants is a request for a recommended ratio between left- and right-hand sides. This is not possible – organizations and situations vary too much.

All you can say with certainty is that a zero score on either side is fatal. Having nothing on the right-hand side, for instance, would mean a complete lack of attention to market and customer matters. Having nothing on the left would mean that no one is 'minding the shop'. Such an organization would collapse, sooner rather than later.

Any score between these extremes must be debated. The purpose of the exercise is been to promote that debate.

Case studies compared

The same kind of analysis can be applied to the detailed case studies shown throughout the book. The blobs in table 7.2 show the relative number of reported results under each heading for each organization and reflect its results, not its expectations. A dash indicates either very few or no results.

These aggregate results suggest where the emphases possibly lay in the minds of these organizations. They are no more than impressionistic. The ratings are subjective – we know of no objective method for doing this – and are not comparable between any two organizations. Adding up or averaging the stars in a column, for instance, would be statistically invalid.

The relative salience of each motive within any of these organizations is unknowable without a detailed briefing, which was neither available to us nor asked for. For example, how do two reported cost reduction results compare with one of increased agility or better customer service? Only the organization itself could answer that sort of question.

No two cases are identical in their profiles but they are similar in their emphasis on internal improvement. There is little apparent attention being paid to doing new things, although plenty on doing existing things differently. Also notable is the lack of compliance as a reason for investing. This will change as legislation and standards such as Sarbanes–Oxley and Basel II begin to bite.

Table 7.2 Case study motives compared

Motive Organization	Cheaper	Better internally	Better externally	New	Compliance	Infrastructure	Image
Canton of Zug	–	●●●	–	●	–	–	●
DVLA	●●	●●●	●●	–	–	●	–
International Truck	●	●●●	●	●	–	–	–
Matáv	●	●●●	●●	–	–	●	–
Morse Group	–	●●●	●●	–	–	●	–
POSCO	●	●●●	●	●	–	●	–
Rabobank	●	●●●	–	–	–	–	–
SDB	●●●	●●●	–	–	–	–	–

These cases seem to us typical. We selected them from about twenty that were on offer at the time.

Changing the focus

It takes time for a group to get used to what it sees in the mirror you have held up to its IT investment history. Once it has done so, it starts to think about remedies. Typical questions are:

- 'Should the emphasis be changed?'
- 'How does this compare with our competitors' investment profile?'
- 'What can we do to get people thinking about outward-looking matters?'

The first question answers itself. The second needs research, probably long overdue. The third signals the start of fresh thinking. There are no pat answers to it, since it involves changes to corporate culture, especially the culture of the IT function, and to investment strategies.

Looking beyond the customary

One simple step is possible, though. This is to suggest to the group that it remove IT investments from the tyranny of ROI thinking. This is one of the most pernicious ideas to be found in commercial computing. It is also inappropriate much of the time.

Paul Strassmann suggests in *Information Payoff* that every method for calculating ROI goes back to the eighteenth century. Capital was then the most important part in organizing the industrial means for production. This antique viewpoint saw all profits created by a company as possible solely because of capital. Labour was a mere commodity. As Strassmann says:

Following this outlook, the financial analysts, the stock market and your shareholders will judge you primarily by your capital efficiency and not by the efficiency of how you utilize your human resources.

Or how you use any other resource come to that, such as your information or processes. A concentration on ROI either ignores all the other six investment motives or else tries to define them in terms of cost saving.

This is ultimately futile. You cannot drive costs to zero, so there is a limit to the gains possible from attacking them. This is not to say that cost control is unimportant – far from it. It is simply to point out that a balanced IT investment portfolio is devoted to other actions as well. As the countryman's saying has it, 'you can't fatten a pig by weighing it'.

Other bases of justification are allowed in almost all other areas of a business. Few organizations do a prolonged ROI analysis of a management development plan, for instance, or the decision to take on a new trading partner. Moreover, much decision-making in business is completely unquantified, especially strategic decisions. The Safeway supermarket in Britain justified a country-wide Extranet on an expected (and realized) better working relationship with trading partners. The value of this to either side was unguessable beforehand.

What commercial organizations need is new and more custom, to drive the business. That will not happen unless money and effort are invested in bringing it in. If companies have not already done so, they must therefore start looking forwards and outwards in their IT investments.

Adopting new justification methods

Why the IT function should have been condemned to use only cost displacement as an argument is one question. Perhaps it is because, in the days of electronic data processing (EDP), companies habitually placed computing under the control of the finance function.

A more important question is why it should willingly continue this monocular way of viewing investment. It is no coincidence that many BPM initiatives begin and remain outside the IT function. A refusal on its part to embrace broader philosophies of investment justification can only contribute to other functions' willingness to go it alone.

If it is allowed to, the IT function must treat justification methods other than cost displacement as legitimate. It must align its practices in this with the rest of the organization. (We recognize that it is not always possible and that the IT function is sometimes the victim rather than the perpetrator here.) If it does not, it risks losing influence over the direction and management of corporate computing resources. A shift from supply-side to demand-side control is happening anyway, fuelled by the rise in Web services and other federated computing arrangements.

The seven investment motives can contribute here. In the same way as you can analyse an investment according to the underlying motives present, so can you justify it. Different motives call for different methods.

There are two points to note. First, no single justification method is applicable to all types of investment. Second, no one method is necessarily sufficient for any individual project. For any particular project, therefore, you might need to use a mixture of methods, expressing some results in financial terms and some not.

Some form of risk assessment is advisable. Government projects in the USA, for example, often include risk-adjusted discounted cash flow (RADCF) analyses of alternative courses of action. This combines calculations of uncertainty about deviations from expected costs or savings with the usual conversion of annual costs to present value (PV). Naturally, this particular method is suitable only where cost-based justification is applicable.

The list in table 7.3 contains some suggestions on justification methods for each of the different motives.

Traditional cost displacement methods have their place, as you can see, but a limited one. ROI arithmetic is valid only for cost containment investments (that is, cheaper).

Efficiency improvements (better internally and better externally) share with cost saving the happy quality that you do not often have to argue for them. Few managers or business people are going to turn down opportunities for either. All people need convincing of is the way in which you propose to achieve the improvements.

The normal method of making a case for efficiency improvement is to show a correlation between the proposed course of action and the expected result. You do this with pilot studies, case histories from similar industries or units, evidence from surveys or projections from existing circumstances. Use argument by analogy, in other words.

Table 7.3 Choosing justification methods

	First-tier motives			
	Cheaper	**Better internally**	**Better externally**	**New**
Primary aim	**Cut or avoid cost**, changing little or nothing	**Improve capability**, changing little or nothing	**Improve performance**, changing little or nothing	**Innovate**, doing things not previously possible, justifiable or imagined
Approach	Custodial	Managerial	Mercantile	Entrepreneurial
Justification method	Traditional cost displacement	1. Show 'proof of concept' (case studies, pilots, surveys, etc.) 2. CBA		Present 'business start-up' case, arguing about ends as well as means

	Second-level motives		
	Compliance	**Infrastructure**	**Image**
Primary aim	**Do what is necessary** to enter or stay in a market or relationship	**Provide foundation**, enabling other systems or investments	**Improve perception by others** – personal, group or corporate
Approach	Consensual	Managerial	Political
Justification method	Yes or no decision	As Better	Anything that works

Cost-benefit analysis (CBA) has its place here. This originated in attempts to put a financial cost on the social impact of airports, housing developments and the like. CBA does not remove subjectivity but does at least parcel it up into discrete areas that can be separately debated, and perhaps agreed on. This would allow people to concur, for example, on the notional value to the organization of faster delivery to the customer or of fewer complaints. You can do so in a similar way to setting sales quotas.

Alternatively, the organization may simply not bother trying to express an improvement, quantifiable or otherwise, in cost terms. The Safeway project mentioned above provides an example of this.

With innovations (new), you have to argue about both the end as well as the means. Change is threatening for many people and they sometimes have to be coaxed into accepting it, even in today's business environment.

Here you take a leaf from the entrepreneur's book. You work up a case for the innovation, much as you would for presentation to a venture capitalist or a banker. Then you show its impracticality by any means other than those that you suggest. Finally, you show a correlation between that method and the proposed innovation, in the same way as for an efficiency case.

You can treat infrastructure investments much as you would better internal investments. The difficulty here lies in the diffuse way in which any benefit may surface. One method is to argue backwards, trying to assess the negative benefit of not making the investment.

Mandatory investments are the simplest to justify. If the organization wants to be 'in the game', whatever the game might be, it must introduce the system or systems required to join it. The only real debate is about the extent of the implementation and the degree of compliance. Does the organization buy a recommended system, for instance, or can it achieve compliance some other way?

There are three likely reasons for investing for appearance:
- as an outcome of some kind of cost-benefit analysis
- as a political transaction
- simply through management *diktat*.

Only the first is worth the effort of formal assessment.

There is a further kind of investment, which we have not dealt with so far. This is the aimless investment, the kind made because 'it seemed like a good idea at the time' or because 'everybody else is doing it'. There are no objectives for it to meet, nor any real deadlines. There is therefore no way of telling when or if the investment has been successful. It is simply a drain on the financial and creative capital of the business.

This is the only approach to investing in BPM for which no case can be made, either literally or figuratively. These investments are debilitating and dispiriting, and they distract from purposeful activities. They are also impossible to monitor. As the old saying puts it – if you don't know where you are going, how will you know when you've got there?

Measuring the results

The methods you use to discover investment outcomes should be consistent with the objectives in mind at the time they were made or approved. Suppose your company invests to improve sales volumes. It would be logical for it to assess those volumes before and afterwards, and to compare the two sets of results.

Life is not always that straightforward. Such simple arithmetical measures are not always applicable, nor is the underlying data so easy to obtain. Where the objective is, say, to improve customer satisfaction or to gain strategic territory, mechanistic data-gathering and manipulation cannot supply the answers.

False alarms and gaps

'Measure what is important; don't make important what you can measure', as Robert McNamara, the US Secretary of State for Defence during the Vietnam War, supposedly said to the chiefs of the US air force. He had discovered that they were using the number of buildings destroyed by bombs as a critical success factor.

Whether this is true or hearsay does not matter; the message is what is important. The managers in many organizations often latch on to something that is easy to measure and use this as an indicator of an investment's performance. There may be little or no logical connection between the activity and the yardstick used to monitor it. Nonetheless, managers use that measure as though it were meaningful. Instead of using what the situation demands, they use whatever falls readily to hand: 'Give a man a hammer and he sees the world as a nail', as the saying goes.

Roger Schmenner, of Indiana University School of Business, publicized the problem in an article in the Australian magazine, *Business Review Weekly*. It reported on some research done in manufacturing industries around the world by the Institute for Management Development (IMD), in Switzerland. Called 'Some Measures of Concern', it begins: 'If it is true that "what gets measured gets managed", then what happens when the measures themselves are suspect?'

The research identified two kinds of suspect measures, which it labelled 'false alarms' and 'gaps'. A false alarm is the use of the wrong measure to motivate managers. They then spend time improving something that has few positive, and perhaps some harmful, consequences for the organization.

Survey respondents reported three such measures that were in common use – machine efficiency, labour efficiency and direct cost reduction. While easy to understand and quantify, they usually gave meaningless results. Their use brings to mind the old joke about the drunk who was discovered looking for his keys in the wrong street. When asked why, he replied: 'Because there's more light here.'

Gaps are the other side of the coin, to mix metaphors. They are failures to use the right measures, so that matters important for the company are neglected. Schmenner gives speed of new product introduction, customer satisfaction and employee involvement as examples of these overlooked but important matters. (To labour the point made earlier, these are the aspects typically ignored by ROI-based approaches to investment.)

Few people would dispute the importance of these outward-looking activities. Sadly, few organizations institute performance measures that relate directly to them. They need only look into how other people have done it and perhaps engage in some lateral thinking. Creative plagiarism is permissible here and to be encouraged.

An organization should know the investment motives for any project beforehand. If it does not, it should at least examine them subsequently. Setting up suitable management measures for existing systems may therefore demand a kind of 'industrial archaeology' to find out what they should be.

If an organization does neither, it cannot be aware of where its false alarms are and where the gaps are. Without this knowledge, it is like trying to measure wind speed with a rain gauge.

The wait for results

Another problem arises with organizations trying to improve their processes. If they are not careful, they could apply unsuitable rules on payback period.

Generally speaking, the difficulty of a change bears directly on the type of benefit likely and the length of time before it makes itself felt. For example, making cost reductions is simple to understand and to carry out. These changes make their way to the bottom line directly and quickly.

In contrast, actions aimed at competitive survival are usually complex and multifactored. The results are not often directly measurable and, in a large organization, can take years to become evident. Even the best BPM software, applied brilliantly, cannot turn a huge enterprise on the proverbial sixpence. Too many other technical, organizational and cultural factors slow down the manoeuvre.

The BT Exact case offers an example. This project is taking place within a large division, itself part of an organization with over 100,000 employees. The timespan for the project is about two years. If something similar were being applied to the whole of BT, we would be talking decades. Indeed, some BT-wide organizational change projects have been in progress at least that long.

Compare BT with iJet. This is a small, and new, business that boasts of having adopted a new business model within hours. Whether that involved extensive restructuring is hard to say but, if true, it is praiseworthy all the same.

Imagine how few projects would gain approval if we were to apply iJet's expectations to BT. Nothing major would get through. Also, there would be a strong temptation for project teams to find some way to fudge the figures to give an unachievable speed of payback. Like many others in their position, the individuals in the team would hope to have moved on elsewhere before they were called to account. Finding and motivating replacements for them would be hard. A happy outcome for the project would be unlikely. Achievable and realistic norms for payback periods avoid this kind of straitjacket.

Snapshot 7.4   Quantum's need for speed

In the days when it was in the disk drive market, Quantum Corporation had hard disks made for it in Japan. It then sold those disks directly to computer makers as well as through distributors. Speed was vital. Quantum therefore gauged the efficiency of its supply chain by measuring what it called 'time of ownership' (TOO). This was the time between Quantum's being invoiced for a drive and itself issuing an invoice to the customer for it. Quantum also took interim measurements, such as time of arrival in goods received. These helped it separate the effects of supply logistics, say, from those of internal handling.

For Quantum, the longer it owned a disk drive, the less value it gained from it. TOO gave it a readily comprehensible and meaningful measure of this and embraced the whole supply chain. It also helped the company identify the costs of product complexity.

This kind of time pressure was not unique to Quantum, nor even to the computer industry. All sectors are having to adopt similar thinking. They can therefore all possibly benefit from automating their supply chains. Ron Factor of Deloitte Consulting put his finger on it when he said: 'Companies do not compete with companies. Instead, supply chains compete with supply chains.'[4]

4 Quoted in *Industrial Distribution* magazine, 1 December 1998.

Case Study 6 POSCO

All the other cases in the book are concerned with application programs that interact with human beings. This case differs in being about the introduction of a process infrastructure.

POSCO, based in South Korea, is the largest steel maker in the world. Part of its success arises from the highly automated production processes it has progressively put in place. The company's turnover is similar to that of Rabobank, the case in chapter 8, but it achieves these results with fewer than a quarter of the number of employees.

Despite this overall efficiency, POSCO was suffering from the slow and disjointed distribution of data internally and poorly integrated supply chain processes. Its chosen solution was to install a 'process backbone' product (see chapters 3 and 4). This links POSCO's mainframe computer to those that run its manufacturing processes and to its trading partner's systems.

Enterprise Application Integration System

Data output from the shop floor systems now get analysed and returned to them immediately, allowing rapid compensatory action where needed. Other benefits include easier management of processes and an improved flow of internal information and to trading partners.

Because this was a system-level project, there was little user involvement. Most of the implementation difficulties were therefore technical in nature. Nonetheless, it was a major project, lasting a year and half and costing US$3.5 million in services and software.

Industry/Sector	Steel manufacture	Location(s)	Republic of Korea
Annual turnover/ income	US$10.6 billion	Number of employees	About 13,000
Type of system	Enterprise process backbone	Supplier and product	Fiorano Business Integration Suite (FioranoBIS)
Number of users	2,000	Time to complete	18 months
Business objectives	• Improve production time and reduce stoppages • Shorten time to market • Minimize rejections of finished product • Optimize use of raw material • Reduce inventory levels and cost • Improve product quality		

	• Increase profits
	• Enable quick and efficient response to business changes
	• Enable business operations to grow
Quantitative results	Transaction rates of over 4 million messages a day
	• Fewer stoppages, faster debugging and shorter time to recovery
	• Drastic reduction in manual intervention
Qualitative results	• Lower maintenance costs
	• Real-time status tracking
	• Use and maintenance of the system simpler
	• System fully adaptable to future needs

Business background

Established in 1968, POSCO is the world's largest and most profitable steel manufacturer. The company makes hot- and cold-rolled steel products, which it sells to the automotive and shipbuilding industries. It can produce 28 million tons of crude steel a year.

POSCO is listed among *Fortune* magazine's Global 500 companies; Morgan Stanley considers it the most sustainable company in the steel industry. The company's annual revenues are US$10.6 billion. A corporate philosophy of adopting leading-edge technologies has allowed it to keep its net margins (12.2 per cent) at more than 3 times that of the industry average (3.9 per cent).

The company's stated values are striving to be the industry leader, valuing innovation and respecting core business and ethical principles. These provide the foundation for its 'mid-range' business objectives of improving competitiveness, securing growth and continuously innovating, especially in business processes.

POSCO wanted to use computers to simplify its steel manufacturing processes and related business tasks. This included connecting the manufacturing systems to its computerized management system, called POSPIA. Launched in 2001, POSPIA uses an Oracle E-Business Suite and integrates POSCO's manufacturing system with its online sales and buying systems. These run across the company's entire value chain.

The problems POSCO wanted to solve included:

Inefficient manufacturing processes It took so long to analyse process data that products had moved off the shop floor before the results were available. Status information about different process steps was all segregated. The information available to different departments was inaccurate and late because data aggregation took so long and was done in batches.

Islands of information systems POSCO's IT infrastructure consists of hundreds of computers. These use varied hardware, run diverse operating systems and differ in the programming languages they can work with. They communicate with either of two transmission protocols and power a wide range of application programs. There is little interworking.

Geographically scattered operations   POSCO has steel mills and offices in multiple locations in South Korea. Data communication was mainly point-to-point. Monitoring and maintaining these systems was difficult, as was developing new application software. POSCO wanted to be able to manage these programs from a central location.

Rigid computing infrastructure   The IT infrastructure was a closed system, resistant to any changes in hardware or software. This resulted in high maintenance and upgrade costs.

System maintenance difficulties   POSCO needed a team of fifteen engineers at each of its steel mills to monitor all its process computers for faults. On average, about four hours would elapse between a problem arising and its detection. The required system would have to permit remote monitoring and troubleshooting, all in real time.

Different data formats   Between them, POSPIA and POSCO's shop floor systems consist of around twenty different combinations of computer and operating system. The files and data stored on them are in multiple formats. These needed to be interchangeable.

Unreliable data exchanging   POSCO's computers are of two kinds, those for managing manufacturing processes and those for managing the business. Links between the two kinds were not reliable. Also, any necessary backup and recovery operations needed the system staff to travel to each machine involved to analyse and compare numerous data logs. These could be more than 200 miles from the place the problem was detected. Remote access of these logs was an important requirement for the new system.

No disaster recovery process   POSCO had no disaster recovery system in operation in all the steelworks. Part of the new system was to ensure round-the-clock continuity of the manufacturing process.

Isolated third-party processes   Trading partners, raw material suppliers, transport providers and others across were using different applications, systems and processes. The new system would need to make it easy – or, in some cases, possible – to sharing information with these systems.

POSCO selected Fiorano Business Integration Suite (FioranoBIS, previously called Tifosi) to meet these and other needs. The software uses Fiorano ESB (enterprise service bus) to transfer and manage messages between diverse systems.

System description

FioranoBIS links to POSCO's company-wide systems backbone, as figure C6.1 shows. The Fiorano software runs alongside an Oracle 9i Application Server on

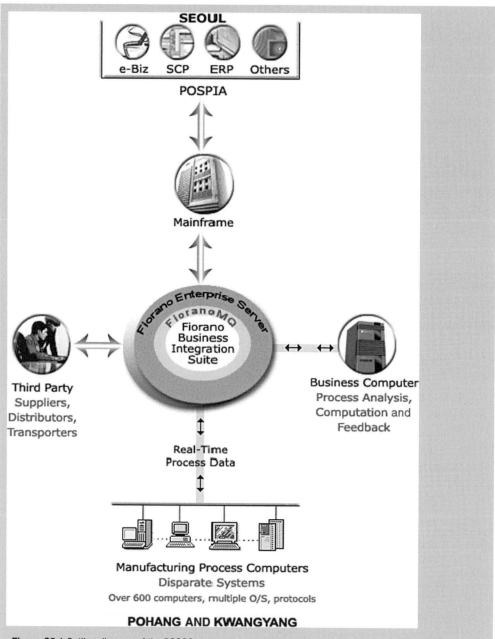

Figure C6.1 Outline diagram of the POSCO system

one of the business computers. It integrates the inputs and outputs of application software running on over 600 computer systems, including IBM and Fujitsu mainframes. These systems manage POSCO's production processes, including some belonging to its trading partners.

Implementation experience

Where and when did the project or system originate?	Seoul, April 2002
How long did implementation take?	18 months
Who did the implementing (own staff, contractors, consultants, etc.)?	Fiorano consultants working with POSCO's team
How much bespoke development was there?	Little – writing a new process data converter and adapting some programs to run on existing computers
Were there any special infrastructure needs?	No
What were the most significant implementation issues and how were they dealt with?	Some, all technical
Who is responsible for the system overall?	The EAI Team (part of the POSCO Process Innovation Group)
How was and is training handled?	Fiorano carried out on-site training
What was and is done to encourage use?	The project was initiated by the Process Innovation Group, so practically no additional encouragement was required
What lessons were learned?	It is important to minimize the chances of human error and to monitor process errors – both directly impact the bottom line; this was made possible with an integrated IT environment

Linking the systems in this way allows manufacturing processes to be monitored, modified and optimized in real time. Any manufacturing inefficiency is now reported to the process computers immediately after computation and analysis by the business computer. The reported data is further used to analyse and optimize the process and remove identified bottlenecks. All the status information is now available for real-time analysis by various departments.

Figure C6.2 shows what a process engineer sees when checking the dynamic links between databases.

Creating the new system and setting it to work took 18 months. Fiorano consultants carried out the work on the Seoul site, with POSCO's EAI team.

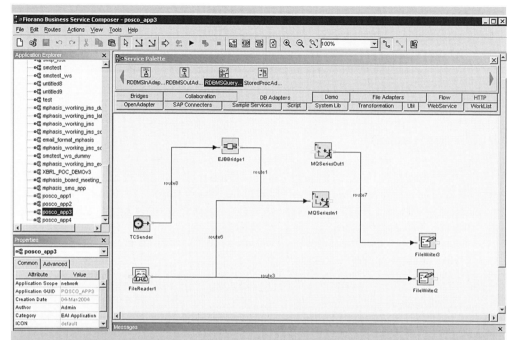

Figure C6.2 User view of part of the Fiorano Event Process Orchestration system

New adapters (converters) were needed to manage some process data. The remaining data interchange needs were met by adjusting the adapters that came with the FioranoBIS software.

The most significant implementation issues were technical in nature, including:

- integrating 500 process computers on the shop floor
- setting up a backup and recovery system; also workstations from one manufacturing plant can now automatically connect to others in case of a disaster
- sequencing messages across 200 process computers, each having its own clock; these messages must be fed in in the sequence they originate.

Fiorano carried out training during various phases of implementation. It will be meeting all future training and support needs at POSCO.

Benefits and user reaction

What has been the reaction of managers and staff?	POSCO's project manager speaks highly of the system and of the support given (see below)
What has been the reaction of customers or trading partners?	All customers and trading partners approve of the new system

What has been the overall cost of the system?	Software and services combined cost US$3.5 million
What have been the main process benefits?	• Real-time exchange of process data • Simplified management of geographically distributed application software • More accurate information, enabling better progress management and planning, order management and reduced time-to-market • Faster troubleshooting • Easy information across POSCO's entire business chain • Setting up a disaster recovery system, leading to continuous and uninterrupted year-round production • The simplified process has enabled greater overall productivity
What have been the main effects on operating style and methods?	Overall operation and infrastructure maintenance is less cumbersome

POSCO moved to a new, open system architecture. Previously, the computer systems were closed and resistant to changes in hardware or software. Maintenance and upgrade costs were high.

Mr J. M. Lee, the head of POSCO's EAI Team and the project manager, says that:

POSCO was looking for a highly scalable solution that could provide seamless data and process integration. We conducted extensive performance, scalability and reliability tests before selecting Fiorano as the preferred vendor. Their solution is an end-to-end infrastructure that integrates back-end systems and includes customers, suppliers and business partners as well. This system has increased efficiencies and yielded significant cost savings. Fiorano ESB's built-in extensibility and standards-based interoperability provides a significant ROI. Fiorano's top-notch technical support made this a logical choice.

POSCO's customers and trading partners have all reacted positively. They previously found it difficult to share information because of the variety of different systems and standards in use. The new system allows them to do so easily.

The EAI project deal included services as well as software.

Some of the major process benefits include:

• Allowing process data to be exchanged in real time for analysis and feedback. This has reduced product reject rates.
• Simplified management of geographically distributed application software. POSCO now has a single comprehensive view of the entire process, manageable from a central location.
• Greater accuracy of information, enabling better progress management, process planning, order management and reduced time-to-market.

- Faster troubleshooting, resulting in faster recovery from problems and errors. It has also brought increased human and machine productivity, reduced down-time and simplified IT management.
- Easier information sharing across POSCO's entire business chain, including third-party suppliers and partners
- Setting up a disaster recovery system, leading to continuous and uninterrupted year-round production.

Users find overall operation and maintenance to be less cumbersome, with a drastic decrease in the need for human intervention.

The future

This new system will allow POSCO to adapt future technologies more easily. It will also be able to use its existing IT infrastructure to realize quick returns in all its future investments.

8 Corporate strategies and approaches for BPM

Management

If business success were simply a matter of intellect and learning, Oxbridge dons would all be as rich as Bill Gates. There must be something else to it. That something is a compound of alertness, opportunism, motivation, leadership, ruthlessness and luck. The same is true corporately. Time has shown over and over again that the smartest companies with the best products do not always survive.

Attempts to nail down the magic ingredients for success are legion. Acres of forest have been sacrificed to feed managers' apparently insatiable need for guidance and reassurance on business strategy. We do not propose to add to the carnage. It is a subject beyond the scope of this book and one on which people with more knowledge than us have written extensively. This chapter instead concentrates on creating a strategy for BPM within your organization's business strategy.

What should a BPM strategy cover?

We show the main elements of a BPM strategy in figure 8.1.

The essence of this is a productive partnership between the abilities offered by technology and the needs expressed by the organization (which for simplicity we have labelled 'business'). The question is not: 'Here is technology. What can we find to do with it?' Nor is it: 'Here is what we are trying to do as a business. What technology is available?' It is, instead: 'How can we harness technology and the business objectives to make something not otherwise possible?'

This symbiotic relationship can occur with other technical resources beside BPM. The central panel in the diagram could be replaced with the names of other technologies and system types. The difference between most of them and BPM is that few other technologies combine its potential strategic importance with its likely impact on human and organizational behaviour.

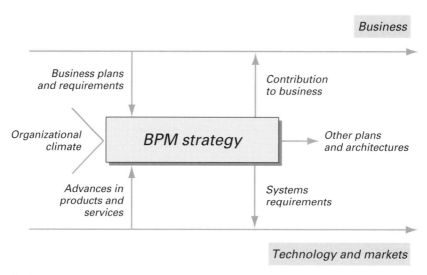

Figure 8.1 Business Process Management strategy and the business

A BPM strategy must do three things:
- contribute to the organization's business strategy
- define the role, content and extent of process management and who has charge of it within the organization
- guide the selection, design, introduction and use of process management systems.

It should do so in ways that:
- support other specific strategies and plans, for information technology and other activities
- take account of the organizational context within which it must operate
- can accommodate planned change and be adaptable to unplanned change
- permit variations to cope with new business demands or new technical possibilities.

We elaborate on these below.

Contribution to business strategy

The BPM strategy must make a clear and direct contribution to the business objectives of the organizational unit (and to those of the overall organization, if originating in a sub-unit). If it can do so visibly – that is, in a documented way – its contribution can be assessed, debated and refined, if needed. It could, for instance, go on the corporate Intranet. This also makes its contribution obvious to other functions within the organization.

We show an idealized and condensed specimen document in figure 8.2.

This document can be as long or as short as local convention allows. Brevity is preferable. If a long document is demanded, it is worthwhile producing a single summary sheet along the lines shown above. It will give readers a ready grasp of what is proposed.

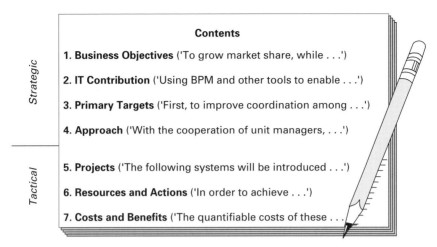

Figure 8.2 Specimen BPM strategy

The flow of the argument is:

- business objectives, where these are known[1]
- the overall contribution that BPM can make, phrased in general but precise terms
- the primary targets of BPM, expressed in terms of functions, processes or organizational units
- approach: this describes the design philosophy, pace, programme leadership and so on.

Those first four items are the strategic elements. They lead on to the tactical elements, which are:

- the projects that will be begun or that were already in progress during the life of the document
- the resources and actions needed for those projects, including money, manpower, brought-in resources, decisions and purchases
- the costs and benefits of those projects.

The document itself, although useful, is not as important as the work that goes into its creation. Two things are needed – that the strategists do their work thoughtfully, and that they do it in collaboration with other functional representatives, especially the users. The strategists must be conscious of three truths, which we have discussed before:

- that computer-readable data forms only a small part of the information within an organization

---

1 Sometimes senior managers do not make business objectives explicit, through oversight or for some other reason. This should not necessarily defeat the BPM strategist. We have known organizations where a computing strategy has been based on assumed business objectives, in a way that allows the higher-level objectives to be inferred. A bolder approach, which we have also seen, is to state those assumed business objectives as 'givens' in the strategy document. Both approaches demand political sensitivity and a certainty that the implied objectives are accurate.

- that computer-based processes can never match the adaptability of those managed by people
- that models are not reality.

At the heart of the planning process is a series of six key questions:

- Where are we now?
- Where do we want to be?
- Why do we want to change? (And will it be worth it?)
- How will we get there?
- What will be the effect of the change (or changes)?
- What will we need in order to make it (or them)?

These, in turn, break down into further series of questions.

Where are we now?

This is an assessment of the current business, organizational and computing situation. Among the questions to be asked at this 'scouting' stage are:

- What are our main existing systems (electronic and human)?
- What is the manpower and organizational situation?
- What are the existing resources for change (people, motives, opportunities)?
- What is the financial and trading picture?
- What problems and needs are evident?
- What plans are in existence or in progress?
- What are the business and other objectives?
- How good are our existing management processes?

That last point is the target of the various Capability Maturity Models (CMMs), which originated in the 1980s. They provide a stereotyped way of assessing the sophistication and formality of computer system projects. These schemes are now being applied outside the IT and manufacturing realms to wider business processes and organizational structures. This is a mixed blessing. CMMs and their relatives have undoubtedly helped bring order and stability to some poorly managed technical operations. Also, as do most models, they provide a common language for debate and decisions. Their disadvantage is that they are as arbitrary and subjective as any assessment scheme. Applied unimaginatively, they can encourage a 'box-ticking' approach that ignores wider contexts.

Where do we want to be?

This is where vision plays its part. The view of the future could be something modest and concrete, such as a defined cut in error rates, or something ambitious and broadly framed. That needs deeper thought.

Such a broader systems strategy might perhaps be to use computers to help increase the number and variety of new products introduced each year. Hewlett-Packard does this. So does Whirlpool Corporation, the American white goods maker. Here is an extract from an examination of its methods by Kathleen Melymuka in *Computerworld* in February 2004:

Whirlpool's approach was to use IT to facilitate innovation much as it has been used to streamline supply chains. The company would re-engineer management processes that slow down innovation and use IT to improve and accelerate the innovation chain from idea to final product. The key was to encourage many low-cost 'stratlets' (or small strategies) rather than a few big-budget projects.

There are two simple questions to ask here:
- Where (or what) do we want to be in the near future? (In five years, say.)
- Where (or what) do we want to be in the distant future? (In twenty-five years, say.)

Short-term ambitions might include:
- making the organization easier for customers to deal with (an explicit aim of Staples, the office products group, and also a driver behind many e-government programmes)
- getting closer to trading partners (what Halliburton, presumably with a straight face, calls 'benevolent entanglement', another way of saying 'lock-in')

A longer-term ambition might be to help the organization move from a reliance on economies of scale to being able to exploit the economies of nimbleness.

Another possibility is to make process competence a competitive weapon. This is not new. Tom Peters, in his 1994 book, *The Tom Peters Seminar*, quotes a marketing executive from the electronics group of 3M: 'Our salespeople don't carry sample kits any more. Now the battle is our flowchart versus their flowchart.' 3M's processes enabled it to engineer a product for a customer's specific needs faster than its competitors could. Now that organizations have access to BPM software, that particular strategic weapon is now within the reach of many more companies than was the case a decade ago.

A further step is to turn process competence into a direct source of income. The American distribution company, DSC Logistics, has done this. It made three important transitions in its thinking and strategy. It:
- made its internal processes explicit to itself and its customers
- converted its processes into a competitive weapon
- turned its expertise in process management into a chargeable service.

From being simply a mover of other people's goods, DSC is now also a supplier of process consultancy and system integration services to its customers. These include Bristol–Myers Squibb, Kellogg's, Monsanto and Yamaha USA.

Why do we want to change?

And will it be worth it? We dealt with these questions in chapter 7.

The question of whether a change is worth it can get tied up in methodological issues if you are not careful. Viewpoint also makes a difference to the assessment of a change. A systems designer might regard a project as a success and an adornment to his *curriculum vitae*, where the users might think of it as a disaster and its managers as only slightly better than no change at all.

Another aspect, which we have mentioned before, is that the speed at which benefits begin to arrive depends on the kind of change. Minor changes, involving little organizational disruption, start to pay their way quickly. Large changes, as typified in business process reengineering, can take decades to pay off. Toyota has been working on its 'revolutionary' production systems for over thirty years.

How will we get there?

There are three main elements to consider in looking at possible routes to the desired future state – business, organizational and technical. The answers to this question contain strategic and tactical elements and, when set down in the strategy document, need to cover such topics as:

- the organization's own definitions of terms such as business process, employee and sale
- the assumptions, uncertainties and programme constraints that apply
- timescales, milestones and decision points
- strategy and project priorities
- proposed systems (try to evaluate at least two possibilities)
- risks, contingencies and remedies
- measures of success
- responsibilities.

What will be the effect of the change or changes?

Each plan or project put forward could usefully be accompanied by an 'organizational impact assessment'. This is modelled on the environmental impact assessments that are usually obligatory for large building or extraction projects. The idea is try to predict the first- and second-order effects of making the proposed system changes, using at least the eight dimensions of organizational character shown in figure 6.3. Another factor that needs to be taken into account is the effects on the individuals concerned – job content, reporting relationships, skill levels, career path, working hours and so on.

What will we need in order to make it or them?

If the proposed changes are to go forward, we need to identify the implications for the area that is benefiting, and for other areas. These will include:

- the human, technical and financial resources required
- decisions required, and from whom and when
- changed procedures, standards and policies
- managing the project or programme

- recruitment, training and support issues
- actions required of the personnel (human resource management or HRM) function, including any payroll changes
- changes to the technical infrastructure
- links with other plans and programmes.

Making changes requires effort and attention not only from the planners but also from the doers. It is unrealistic to expect an organizational unit that is undergoing change to be able to perform at its usual levels. It is equally unreasonable to expect a unit to make changes in structure or working practices when the people in it are fully stretched in dealing with day-to-day work. You cannot stop the business while the planning process or an implementation project goes on.

In the early stages, at least, the strategist or strategists should be aiming for the broad picture at the higher levels. Fine detail is for tactical issues. Although the process should not be rushed, neither should it be agonized over. It is better to be roughly right quickly than exactly right slowly.

Not all situations will require such a thoroughgoing approach. Although purists might argue that all design efforts should involve these broader considerations, even at the tactical or operational level this is a counsel of perfection. Designers must fit the means to the circumstances. If a 'quick and dirty' project is all that resources permit, a shorter version of the strategy process will have to suffice. So long as the participants in the exercise recognize it for what it is, the potential for damage will be limited. It is when such short-termism gets presented as long-term strategy that real damage results.

Strategy as a collaborative activity

'Well-informed employees are our best ambassador.' So, some years ago, said an IT manager at Welsh Water, a British utility. Every employee should be aware of what his organization does, what its objectives are and where it stands in relation to its competition.[2] Those same employees should at least be invited and helped to participate in planning how the organization is to reach those objectives. Sometimes they will expect to.

Employee participation in planning is not just the preserve of small companies. It is accepted practice in organizations like the John Lewis Partnership in Britain (59,000 employees at the time of writing), Mondragón Corporación Cooperativa in Spain (over 68,000 employees at the end of 2003) and, in the USA, W. L. Gore & Associates, the makers of Gore-Tex fabrics (annual revenues of over $1.2 billion).

---

2 A similar principle applies to strategy maps. These stem from the work of David Nolan and others at the Balanced Scorecard Institute. Since its first publication in 1992, the Balanced Scorecard has been helpful in getting managers to think of matters other than financial ones when planning and controlling. The strategy map provides a pictorial summary of an organization's scorecard. In organizations run on more inclusive lines, the Balanced Scorecard and the resulting map will probably be overly hierarchical.

It reaches its apogee in Semco SA in Brazil, whose employees, over 3,000 of them, exercise a level of control most managers would be uncomfortable with.[3]

Whatever kind of organization yours is, strategy is not something only specialists do, in ivory towers. It should be done where it applies, by those to whom it applies. This principle applies within limits, of course. A planning meeting of the whole workforce would be unmanageable. One consequence of this is that there should be as many strategies as there are separately identifiable needs, possibly more.

If these various strategies are not to diverge from or conflict with each other, they should not be created in isolation from each other. Also, there needs to be some elasticity within each separate strategy. Without some 'give', they will not work together. Strategizing should thus be a collective and coordinated set of activities, out of which are produced a series of interrelated and adaptable plans. Cascading tablets of stone cause only headaches.

Role, content and management of BPM

We have dealt at length in earlier chapters with the role and content of BPM. Its management is often split, with process owners having charge of its application, rules definitions, performance standards and day-to-day running. The IT function typically has responsibility for the system's service levels and for capacity planning, network management, technical training, data policy and routine support.

The subject of central versus local control is a perennial focus of attention. Some people see the centre as a value destroyer, not a value creator. For example, divisionalized organizations often pay a levy to the centre. In return for this, as they sometimes see it, the operating units get alien and irrelevant policies imposed on them, have their funding trimmed, their ideas stifled or plagiarized and their plans interfered with.

This viewpoint is understandable and, in some cases, even accurate. These inhibiting tendencies have been attacked in companies such as the Kyocera Corporation in Japan and Oticon A/S in Denmark, who have adopted what are still regarded as radical approaches to operation and structure. Yet, even in these dispersed and devolved arrangements, the centre retains control of overall strategy and of standards. Maintenance of the company culture also stems from the centre, in establishing an all-embracing and internally consistent set of expectations, controls, resources and attitudes.

---

3 Ricardo Semler, Semco's CEO, has written two books – *Maverick!* and *The Seven-Day Weekend* – that describe in detail how its human and business systems work. The books are spaced ten years apart, showing how the original ideas have fared. Also, contrary to some critics' expectations, the gap proves that an organization adopting these ideas can survive and prosper.

Similar problems and remedies apply to the management of IT. Unless the central IT function conspicuously acts as an enabler and a facilitator, its influence will often be seen as negative, even when it is not. In these days of cheap computing, operating divisions find it easy to go their own way regardless of the effect this may have on any central strategy.

The modern user is wary of attempts by central computer departments to regain control over corporate computing. Having tasted freedom, through the introduction of the PC, he will not willingly let it go. He welcomes advice but not restriction.

Also, these days, the computer user has opinions. He has become accustomed to the fast-system responses, graphical user interfaces and local printing that personal computing brings. He is scornful of traditional data processing approaches to service delivery and usability, and has sufficient power to back up that disdain. Unlike earlier generations of workflow automation, which were aimed almost exclusively at routine clerical tasks, BPM is as likely to be applied to the work of people in a managerial or professional positions. These people have the power to enforce their point of view.

In consequence, many local systems have been specified by operational units, paid for by them and provided by external suppliers. Often, the computer department's role, if has had any, has been that of specifying standards or interfaces. On some occasions, the central IT function has not known until afterwards that the software was going in. In some organizations, the only corporate IT standards that remain may be those that ensure connections among users and systems.

Strategic IT planning should mainly be done at the level of the business unit. The most important tasks of the corporate IT director or CIO are to advise business units on their strategic IT plans and to influence corporate strategy to take full advantage of IT.

Not all senior IT managers are willing to do this; they are more 'hands on'. You are unlikely to be able to change this state of affairs but it helps to recognize the kind of IT manager you are dealing with.

Support of other strategies

Any BPM programme will affect, and be affected by, other information technology strategies and policies. These will include communications infrastructure, data management, security policies, storage architecture and programming standards.

It will also affect information management plans (sometimes grandly called 'knowledge management'). Not every organization has this as a separate discipline. It is mostly found in large, information-intensive organizations, such as insurance, pharmaceuticals and oil companies and in academia and government. For any organization, BPM will probably introduce a dynamism of information flow that did not exist before. If there is an information management function, it should also involve itself in setting the BPM strategy. Data definitions may also need unifying.

There are other plans and policies that need to be taken into account. These can include manpower plans, security policies, budgeting processes, intellectual property policies, marketing plans, facilities plans, quality management programmes and capital plans. Getting the early involvement of their planners is prudent, if only to avoid later conflict and possible backtracking. More positively, they will have ideas and know of other activities that might improve the BPM strategy.

Organizational context

We dealt with this in detail in chapter 6. Even where the process or business planner feels confident of his knowledge and ability in these matters, it is worthwhile involving at least one organizational specialist in the strategy process. His second viewpoint and his separation from the detail of the overall strategy will provide, as a minimum, a sanity check on proposed changes.

Designing for change

One of the favourite areas for bought-in consultancy services is that of change management. Its basis is that managers need help in understanding the principles behind making successful changes and training in how to apply those principles in a variety of circumstances. On the whole, this is largely unexceptionable. The problem arises when people regard change management as a subject separate from day-to-day management and separate from other aspects of introducing computer systems. This can lead to its being treated as something to be applied to the situation like a kind of ointment, instead of something integral to it. Like planning, change management should be practised by those who must live with its results.

BPM in the medium-to-large organization must be based on an architected approach. Reliance on one supplier can be risky, especially if that supplier is not in the first rank. Similarly, reliance on one technology is hazardous; new technologies and new ways of packaging existing technologies become available almost every week. An architecture releases the customer from reliance on one supplier or one technology, and exposes the areas of certainty and those of volatility.

The contribution of an architecture

Architecture can provide other benefits. In their book, *Paradigm Shift*, Don Tapscott and Art Caston examine the reasons for having a systems architecture. We show these in modified form in the list below, discussing them immediately afterwards.

A systems architecture:
- begins with an outline plan, like that shown in figure 8.2
- needs an architecture team to discuss, create and manage it
- applies frameworks and principles

- provides standards and guidelines
- resolves interrelationships
- defines and arranges components
- simplifies design and purchasing
- permits minor deviations.

The starting point is the business plan. That defines the objectives, resources and constraints that will apply. From it stems the rest of the work.

Three points deserve attention. The first is that this is itself a process and one that must be repeated. This does not mean that you should throw the architecture open for complete re-examination every few months. That would be wasteful and demoralizing. It does, though, mean you should review it as the organization's systems and their environment change.

The second point that an often unrecognized contribution of an architecture is to permit minor deviations. It does so primarily through a form of variety reduction, by defining what is minor. Should the central IT function worry if a branch office adopts Sun OpenOffice instead of Microsoft Office, for example?

Without an architecture, a map of the future, one cannot know whether such an action is significant. This is typically the case in organizations that define systems activities in terms of compliance with rules rather than the pursuit of objectives. A policy is not a plan. In these days of shrinking central power and its replacement by influence, such an approach is unhelpful. The role of the central group is to assist downwards, influence sideways and advocate upwards.

Finally, the list can apply not just to BPM, or even to IT generally, but to most technical areas within an organization. Its elements are equally relevant to engineering, manufacturing, design and facilities planning, for instance. Many also apply to activities that are not based on technology, such as marketing, purchasing, training and manpower planning.

It follows, therefore, that these various architectures need to be kept in step with each other, in the same way as the various associated strategies should. The assumptions on which they are based should be mutually consistent, the objectives they are pursuing should be compatible with each other and the timescales over which they are operating should be in harmony. And the only way to achieve all this is by regular and meaningful communication among the architects.

Dealing with conflict

BPM should be built on cooperation and collaboration, but these are not always present. Only the most unrealistic person would imagine that these qualities are to be universally found or that they are constant in their effect. The reality of organizational life is that individuals and groups do not always trust and respect each other. People are prey to all sorts of emotions, not all of which are noble or rationally explicable.

There will be times when cooperation is absent and hostility and suspicion are the norm. What does the BPM strategist do in such circumstances? Give up, proceed as if nothing were wrong, try to achieve harmony or do something else?

If the problem is local or short-term, the strategic impact of these negative forces is likely to be insignificant. That is not to minimize their tactical or operational effects. Dealing with human considerations is a primary requirement in introducing any new system but it should be dealt with locally. Should local problems continually recur, they may be symptoms of a more serious, underlying condition.

If these interpersonal or intergroup hostilities are endemic throughout the organization, then the strategist must accept them as fact and base his approach on their existence. Instead of the expected trio of coordination, cooperation and collaboration, he must instead deal with conflict, compliance and competition. This limits the kind of system you can introduce and the way it will be used.

For BPM strategists in conflict-ridden organizations, who perhaps do not enjoy much direct power, life is very much a case of cutting one's suit according to one's cloth. The kind of system, and the way it is used, will most likely be required to provide clear-cut financial benefits, and quickly. The investment motives for the systems will incline to the left side of the 'virtuous circle' we discussed in chapter 7. They will be inward-looking, which will take away much of the point of engaging in BPM.

Making real change

Before you disturb the system in any way, watch how it behaves . . . Starting with the behavior of the system forces you to focus on facts, not theories. It keeps you from falling too quickly into your own beliefs or misconceptions, or those of others . . . [It also] directs one's thoughts to dynamic, not static analysis – not only to 'what's wrong?' but also to 'how did we get there?' and 'what behavior modes are possible?' and 'if we don't change direction, where are we going to end up?'
And finally, starting with history discourages the common and distracting tendency we all have to define a problem not by the system's actual behavior, but by the lack of our favorite solution.

This quotation is from the systems writer, Donella Meadows.[4] Dr Meadows also compiled a widely reproduced list of places where intervening in a system would make a difference. The version of the list we show below is based on her 1990 paper, 'Places to Intervene in a System'. The higher levels in the hierarchy subsume the lower. Intervening at a lower level is of limited value if the higher levels are left unchanged:

1. The power to transcend underlying models, such as by challenging assumptions. ('If I were you, I wouldn't start from here.')

---

4 It is from her article, 'Dancing With Systems'. Despite the waffy title of the piece (and its introduction), this is a practical outline from a widely experienced systems thinker and consultant.

2. The worldview out of which the system arises – the unstated assumptions, for instance.
3. The goal of the system.
4. The power to change the system's structure. (Who is allowed to do so?)
5. The rules of the system.
6. The structure of information flow.
7. The strength of positive feedback loops. (These reinforce and amplify themselves.)
8. The strength of negative feedback loops. (These moderate and constrain themselves.)
9. The length of delays.
10. The structure of material stocks and flows.
11. The size of buffers, such as inventory and work queues, relative to their flows.
12. The parameters and numbers used, such as measures of performance.

These are not rules, and the list is open to debate. As Dr Meadows says of the list, '[It] is tentative and its order is slithery. Every item has exceptions that can move it up or down in the order of leverage.'

The attention each of these points gets is often in inverse proportion to its effectiveness. People argue vehemently over measures, for example, but hardly ever concern themselves with items 1–4, say. These are the most potent determinants of a system's behaviour but the slowest to get changed, as our diagram in figure 6.3 and its following text also highlight.

In practical terms, the message is to act on items as high in the list as you can reach if you want to make a real change. Reading Dr Meadows' paper and other works referring to it will give you a fuller idea of the principles involved.

Managing your BPM programme

In many respects, managing a BPM programme is the same in its essentials as managing any other cross-functional initiative. It calls for political skill, a willingness to communicate and keep communicating, persistence, a sense of direction, plenty of patience, optimistic endurance, low cunning and some (but not a great deal of) technical knowledge.

Although BPM differs in important details from TQM and BPR, for instance, there is much you can learn from the experiences – negative and positive – that people have gained in both activities. Human aspects change little compared with technical ones. There is plenty of material on TQM and BPR to be found in books and on the Internet (see the suggested reading list at the end of the book). Bear in mind the differences among the three activities (see table 2.1).

There is also useful evidence contained in the case studies and other user material in this book. In addition, if possible, compare notes with someone who has been involved with either a successful TQM programme or a sensitively applied BPR programme. BPM is not a re-run of BPR, so be guided by reports of what actually happened, not what some propagandist said should happen.

Remember also and always that there is no one best way. What has worked well in one place may not necessarily do so in another, so you must select your evidence with care.

Objectives, demands, pressures, resources, history and organizational context combine to make every programme different. Especially beware anyone who insists that some particular course of action represents 'best practice'. This is one-best-way thinking wearing an official badge (see chapter 2).

Two distinctive differences

There are two aspects in which BPM does differ from many other corporate programmes. The first is in the need to get people, especially senior managers, to think in terms of systems and processes. By these we mean organizational systems, not just computer-based ones, and processes that extend to the organization's boundaries and beyond.

Naturally, before you can persuade and educate other people in these topics, you need to feel convinced and comfortable yourself. We hope that this book has helped with this, as will the reading material we list at the end of it.

Even then, reading is no substitute for experience. You may perhaps want to take advantage of relevant local or online educational programmes, especially those with some practical content. Exchange visits are helpful, too, possibly through your professional institute or chamber of commerce.

Alternatively, or as well, you may wish to arrange some experience transfer by engaging a consultant specializing in systems thinking. This person need not be *au fait* with BPM or even with computer-based systems. He should certainly not be wrapped up in computer-based thinking.[5] His task is help you and your colleagues elevate your thinking to embrace abstract models of the kind Stafford Beer created (see chapter 4).

After all this, there is no substitute for learning by doing. Since you cannot make pretend changes to a real system, it makes sense to make real changes but where the least damage is likely if things do not go as planned. In other words, do pilot systems. Chose your pilot system carefully. To have the best chance of success, it should:

- be relatively simple and small-scale
- have enthusiastic managers and users
- have easily measurable results (before and after)

---

5 Systems integrators are unlikely to be of help here because they are usually technologists first and general systems thinkers second, if at all. Also, their habit is to keep their craft knowledge to themselves, as it is a saleable commodity.

- be where subsequent publicity and dissemination are easy
- conversely, be where lack of success can be passed over easily
- have a low cost of failure
- be easily translated to other areas.

Unless you are managing a computer department and are concerned with its systems, it is seldom wise to do pilots with computer specialists. They are fond of doing so themselves, as it happens, but almost always concentrate on the technical aspects of whatever they are trying out. Also, because computer folk are not like other employees, you cannot generalize from their experience to that of the rest of the organization. They are, for example, often tolerant of user interface character- istics that other users would complain loudly about.

Getting the support of the IT function

This is the other main area of difference from most other corporate programmes and initiatives. BPM is impossible without the adoption of the relevant software. You therefore need access to computer expertise – your own, your organization's or somebody else's.

The question is: what kind of expertise? For a sustainably successful BPM program, systems design should be built around accepting and exploiting:

- human characteristics
- differing viewpoints
- variability in processes
- constant change.

Ideally, organizations should take these four factors into account in the methods they use for planning, analysing, designing, implementing, managing and main- taining all computer-based or computer-aided systems, not just BPM.

Radovan Janecek of Systinet, a software supplier, comes to a similar conclusion when considering the software infrastructure needed to enable BPM. In his view, it should:

- facilitate processes in more flexible way
- make humans first-class citizens of machine-operated processes
- give people direct visibility and control over corporate data flows at any time
- allow them to engage in *ad hoc* improvisation
- support evolution of the process definition to react to a changing environment.[6]

Software that can do all these things is of limited use unless the people who design systems that use it follow the same precepts.

It could be that your organization's existing methods embrace these principles. If so, that is to the good. If not, then you must recognize the inertia behind your computer department's preferred way ('methodology') of designing computer-based

---

6 From his 'blog' (Web log), *Nothing Impersonal*, 19 October 2003.

systems. Systems managers may not be willing to change that method, unless it is proving unsatisfactory on other grounds. The organization's financial and practical investment in it will be probably huge. Also, there will be people in the computer function with a large personal investment in its retention.

At this point, negotiation and judgement come in. These are the choices in front of you:

- Try to soldier on with the existing corporate methods and hope they will not prove constraining
- Try to persuade the IT function to follow your lead
- Go ahead on your own and leave the IT department out of it
- Use your own methods where you have control; for those aspects the IT function controls, try to interwork with the existing corporate methods.

We cannot advise you on which you will have the greatest success with, as we do not know your organization or your situation. We can, though, tell you that the likelihood of the first succeeding is the smallest of the four.

The second is hard, too, but potentially the most rewarding. Unless the IT function is already committed to user-based design, you will be confronted with a list of reasons why what you propose is not possible, not affordable, inconveniently timed, technically incompatible or simply wrong. If you do not have the power to direct that the IT function look into this in an open-minded way, then you must resort to persuasion, subterfuge and, if necessary, deceit. Politics, in other words.

Traditional approaches to systems design assume that:

- the objectives of the system are fixed and universally accepted
- users have a perfect understanding and recall of what they do and why
- the analyst makes a full, accurate and unbiased record of what users tell and show him
- the programmer faithfully translates the resulting specification into a set of computer instructions
- users can readily learn how to use the resulting program
- the demands it was designed to meet have not changed since the start of the project.

How many of these assumptions are likely to be valid, especially in today's business setting? Exactly.

Despite this, the literature on BPM is full of system design diagrams showing tidy, concise, self-contained and one-way routes to software success. Looping, revisions, uncertainty and overlaps are all absent, as is any more than token user involvement. This is mainframe thinking with a new coat of paint.

A further point is that traditional design methods guarantee that users' learning will always lag behind design, usually by a long way. By the time users have got to grips with the system and properly understood its workings, the system will have been signed off and, probably, its design team disbanded. If users detect flaws or see areas for improvement, these go down on the list of 'maintenance' items. This is not maintenance, of course, merely making the program do what it was supposed to in the first place.

Modern design techniques allow users and designers to work together at all stages through the development of software.[7] Designers repeatedly give users the equivalent of rough drafts of the new system, to experiment with and report back on.

The result is a better system:

- users' learning proceeds at the same pace as the design process, and sometimes in advance of it
- their stated aims and needs are dealt with as completely as possible
- they can clarify in their minds and then make explicit their previously unstated aims and needs (people are often unaware of exactly what they do in a work process and why)
- they and designers can immediately test whether expressed needs are an accurate and 'robust' reflection of the reality they know
- design is faster
- maintenance is reduced.

The IT function in your organization might already be considering BPM software to ease their system integration and maintenance burden. If so, it is a good time to start talking to your computer experts about adopting new design techniques to help the organization make the most of the investment. If not, you could start sowing the seeds of both, based on what we have said over the preceding few hundred pages. Should neither course appeal to you, and you want to go it alone, you should at least by now have a good idea of the task in front of you and how it relates to your organization's aims.

Whichever course you choose, there is one other ingredient you will need and that is luck. Every project, programme and organization needs that. We wish you it in abundance.

Business process musings

Petra in 't Veld-Brown is Programme & Process Manager at St Regis Paper Company Ltd. Based in Taplow, Berkshire, the company is Britain's largest paper and board manufacturer. Petra is responsible for developing and managing the company's strategic IT programmes and business process frameworks. This includes selecting and defining a process method for the company. Outside work, she is co-leader of the BPM Group's special interest group on BPM methods.

---

7 These techniques go under various names, such as extreme programming (or XP), Rational Unified Process, Dynamic Systems Development Method and agile software development.

Before interviewing Petra, we sent her three questions to mull over:

- *What are the problems that BPM is there to solve?*
- *Does it look likely to be able to do so?*
- *What are the obstacles to this?*

The conversation ranged wider than this, of course. Here are our notes of it.

On the need for process management

(Petra) What problems is it there to solve? All of them! BPM, and process thinking, is like computer security – you don't bolt it on afterwards. It's integral to running the business. If you understand what you are trying to achieve through and with the business, process management is intrinsic.

How to go about it

(Petra) You need to have a strategy and business plan in place before you start BPM. If nothing else, this makes sure that what you are doing will be part of the future of the organization. Otherwise, you end up with a sticking-plaster approach.

Sometimes you need to do process improvement by stealth, building it into other plans and policies.

Bring the organization down to its core activities, so you can ensure you're doing the right things, in the right way and so on.

Going middle-out is as important, and as widespread a need, as top-down or bottom-up.

Problems with BPM

(Petra) Organizations use IT to manage processes whose detailed workings people cannot keep in their heads any more. The weakness comes from treating those processes as 'black boxes', whose contents we take for granted.

Too many corporate processes are not explicit and visible. Different people do different parts, invisibly and unknown to others and unknowing of them. It's like a team game of pinning the tail on the donkey.

People believe in BPM technology as though it's a panacea. It doesn't tell you what to automate, or how.

In the wider picture, people view IT today the way they regarded engineering a century or two ago. They still haven't adjusted to it in their thinking. It's human mental adaptability that sets the pace, our ability to absorb these new abilities and ways of working.

The importance of people

(Petra) The biggest obstacle is that the people doing process change don't understand the people who perform those processes. Worse, they do not even recognize that this is the problem. There is a need for empathy from process and system designers.

Financial drivers and timetables are the biggest disincentive to involving people. Managers are often more focused on meeting the budget than working towards the strategy. The rush to shareholder value gets in the way of broader considerations. This can result in bullying. Too often you hear expressions like: 'They will just have to do what we tell them' and 'If they can't adapt, we'll get someone in who can.'

Every human problem boils down to communication. You must make people feel part of what's going on. They can be embarrassed or ashamed at what they feel is their lack of knowledge. You're better off telling them you've done nothing than not telling them you've done something.

Aggrieved people don't or haven't had the chance to be heard. They need to have what happened to them recognized. It's insulting and humiliating otherwise.

Cultural differences matter, too. In the Netherlands [Petra's home country], for instance, people can speak to senior managers as though they were on the same level as them. They can't in the UK.

Case Study 7 Rabofacet

Rabobank is the main retail and merchant bank in the Netherlands. It has a federal structure, made up of a central core and nearly 350 local banks.

This structure, while good for customer and community relations, was bringing difficulties in dealing with customer requests. These could arrive along any of at least three routes, online and offline, and be replied to the same way. Responses to these queries were therefore not all meeting internal standards of speed and quality.

The bank set up a central operation to handle queries and requests, which it put under the control of Rabofacet, part of the core. Rabofacet introduced process management software when creating the new fulfilment centre. The software runs the monitoring and control processes and links to existing customer relationship and product systems. It also links to an email server and to the bank's printing factory.

Turn-round times for customers' queries and requests have been halved, even though their volume has increased to around 20,000 a month. This has been achieved with fewer staff. While they now have better access to customers' details, it seems that only the centre's managers get a picture of the whole process and its results. Local banks now entrust this aspect of their back-office activities to the fulfilment centre. The system will embrace new processes and functions in future.

Industry/Sector	Banking	Location(s)	The Netherlands
Annual turnover/income	8.6 billion euros in 2002 (approx. US$11 billion)	Number of employees	57,500
Type of system	Workflow and BPM	Supplier and product	Staffware
Number of users	10	Time to complete	9 months
Business objectives	• To manage an increasing volume of customer communications in a timely and cost-effective manner • To manage a wide range of processes, successfully		
Quantitative results	• 20 sales processes and 30 service processes supported • Handles more than 20,000 customer requests a month • Dramatic reduction in handling times, e.g. from at least 10 minutes to less than 30 seconds • Turn-round time for customers cut from up to 7 working days to 2–3 days • Only 5 back-office employees are needed to handle all the above		
Qualitative results	• The complex back-office process can now be developed and managed uniformly and unambiguously • Integrated management information is available, presented uniformly, on each channel, product and bank • Most changes to the process can be made directly while in production and without having to change the program source code		

- Uncoupled user interaction means that employees can continue working if any contributing system is temporarily unavailable
- The fulfilment centre gets an integrated picture of work outstanding, allowing it to prioritize and divide work efficiently
- The centre can also now oversee the whole process, including monitoring progress in other departments
- Employees with less experience can now handle these complex processes.

Business background

Rabobank Group is the dominant Dutch bank and the sixteenth largest bank in Europe. At June 2003, it managed funds worth over 175 billion euros (roughly US$225 billion). The bank has over 9 million clients and serves more than half the Dutch population and businesses. It is the country's market leader in mortgages and private savings, and in banking for agricultural organizations small and medium-sized enterprises (SMEs).

The group's structure is federal, comprising over 340 local banks linked to a central organization. These local Rabobanks manage more than 1,400 offices in the Netherlands. The bank also has nearly 200 offices in thirty-four other countries. It employs over 57,000 people.

Rabofacet provides IT services, purchasing, logistics and facilities management to Rabobank Group entities. It employs roughly 2,500 people at sites in Zeist, Utrecht, Eindhoven, Tilburg and Best. Rabofacet manages an IT infrastructure that includes 3,000 ATMs and 1,700 branches. It also centrally processes business transactions and provides workstation support for about 40,000 employees.

Project background

Customers send in many different requests and queries to the bank. These include requests for sales and service processes and for information. Requests and queries arrive through different channels, such as the Internet, call-centres and, for direct mail coupons, the postal service.

In 2000, Rabobank realized that local banks were not handling queries via 'virtual' channels in a way that met internal quality standards. Also, customers using direct channels such as the Internet expected a faster response than with other methods. Rabobank often invites its customers to visit their local branch offices, but some customers prefer remote banking. Rabobank decided to set up the fulfilment project to improve the quality and efficiency of handling such queries. The main change would be to route queries through a centralized fulfilment centre, handling the vast majority direct. It put the project under the control of the existing Rabofacet Operations Centre in Best in the Netherlands.

Rabofacet created a workflow management system for this. Called TRACK, it uses Staffware software and is the basis of a sophisticated back-office fulfilment centre.

The eventual aim was automatic fulfilment, with staff handling front-office processes and full automation of back-office systems. 'Nevertheless, there are several reasons one never reaches this Utopia', says Henk van Dijkhuizen, the responsible project manager at Rabofacet. 'New products are launched, products change, and ICT systems are usually not able to cope with the dynamic of marketing. Therefore, the fulfilment centre is a perfect migration platform. It allows us to fall back to a 90 per cent optimized process, instead of waiting to reach the ultimate 100 per cent.'

System description

The main features of TRACK are:

1. Multi-channel back office

   By clearly separating front and back office, Rabofacet could set up the back-office process in a uniform and channel-independent manner. All the various communication channels connect to one integrated back office.

2. Adopting a process-oriented approach

   This was achieved through using Staffware's BPM software. Also, the division between front and back office makes it possible to sub-divide complex processes into process steps. For each step, the system's designers could decide whether to handle it automatically by system integration (see below) or to have employees enter the data.

3. High level of application integration

   The TRACK Fulfilment System links to more than eleven Rabo systems, mostly through IBM's MQ Series backbone software. These include:

   - Customers system and customer contacts system. This allows the bank's branches to stay informed in real time about the progress of queries. TRACK also produces detailed, real-time management information.
   - Product databases, such as those for saving accounts, insurance, transaction approval and virtual banking. This lets most process steps be handled automatically.
   - The bank's printing factory in Rabofacet. Through this, contracts and other documents for the customer are automatically printed and processed.

- There is, in addition, integration with an email server. Where the business rules defined in the workflow say that a customer request needs local attention, TRACK automatically forwards the request to the local bank.

Henk van Dijkhuizen says, 'Handling processes in the back-offices under the control of the Staffware BPM software means we don't need to reach the highest quality standards in all our back-office systems. In fact, the temporary failure of a back-office system is never a reason to lose productivity in the fulfilment centre. Staffware will try to contact the back-office system later. The employee is not involved before all relevant information is collected.'

Figure C7.1 shows how the fulfilment centre uses the process software to keep track of the documents needed to decide the loan possibilities for a customer.

Implementation experience

Where and when did the project or system originate?	The project was launched in January 2001. The TRACK system began working on 1 September 2001
How long did implementation take?	From development to implementation took 6 months; since then, there has been a new release of TRACK, with new or improved processes, about every 3 months
Who did the implementing (own staff, contractors, consultants, etc.)?	Rabobank did the implementation A Staffware consultant joined the project team for the first release
How much bespoke development was there?	A significant amount; development of the first release took about 3 months
Were there any special infrastructure needs?	Yes: Rabobank needed to enable its infrastructure to allow a central organization (Rabofacet) to use systems developed for local banks; IBM MQ Series middleware was introduced for the task, in 2001
What were the most significant implementation issues and how were they dealt with?	The greatest difficulty lay in producing a detailed specification for the process; the amount of detail made testing a major issue, as did the extent of integration with other application programs
How was and is training handled?	By using TRACK, most employees do not need to understand the full context of the processes so general training investments are low; for complex processes, such as loans and insurance, training of several days with process specialists was organized
What was and is done to encourage use?	No significant investments were required
What lessons were learned?	<ul><li>'The devil is in the details': detailed specifications are vital</li><li>Testing is complex and can be done only by thoroughly planning the process</li><li>For delivering quality, the total chain in the process needs to be managed actively</li></ul>

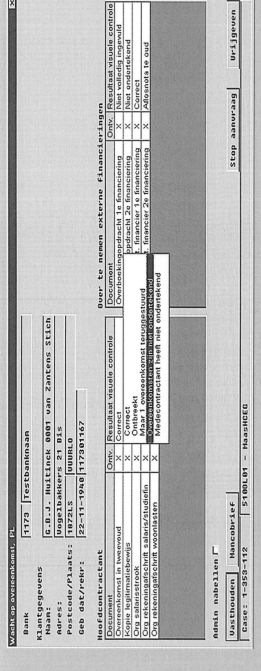

Figure C7.1 User screen of the Rabofacet fulfilment system

In centralising the processes, the fulfilment project learned that despite the efficiency reached centrally, some customer requests are better handled at the local branch. Henk van Dijkhuizen says: 'We are always looking for the best balance between efficiency of central and automatic fulfilment, combined with personal service and advice through our local banks. We don't believe in an isolated central bank approach. Instead, the fulfilment centre must enable the local bank to focus on the personal advice to Rabobank customers.'

Benefits and user reaction

What has been the reaction of managers and staff?	Enthusiastic. Managers and staff are pleased to have a single environment that displays their work position and that of their colleagues. Operational management and marketing are aided by the sophisticated and integrated management information available.
What has been the reaction of customers or trading partners?	• Customers don't see the system but are happy when they can do their banking activities 'at home'. • Although local banks hesitated at first, they are now convinced that they should organize their back-office activities centrally.
What has been the overall cost of the system?	Company confidential.
What have been the main process benefits?	• Can now process a greater volume of customer requests and queries with no increase in staff • Fulfilment centre staff now have full visibility of customer details, quickly and clearly • Managers have better visibility of processes and results
What have been the main effects on operating style and methods?	• The (temporary) unavailability of a product system does not adversely affect employee productivity • The fulfilment centre now has an integrated picture of the total amount of work outstanding • The fulfilment centre can now oversee the whole process • Less experienced employees can carry out more complex processes

Business benefits

The integrated workflow environment Staffware provides has made it possible to reduce the turnaround time for customers from 3–7 days to 2–3 working days. Customers no longer need to go to the bank for their requests to be actioned.

Henk van Dijkhuizen says: 'Synergy benefits have been achieved through the multi-channel approach. Innovative power and manageability have both been substantially improved. The often complex back-office process can now be developed and managed uniformly and unambiguously.'

The multi-channel, process-oriented approach has resulted in integral management information being available for each channel, product and bank in a uniform manner.

Using workflow automation has allowed the process to be adapted in a flexible fashion. Most changes to the process can be carried out directly in production without needing to change the program source code.

Operational benefits

Unlinking system integration from user interaction in the workflow design means that any temporary unavailability of a product system does not adversely affect employee productivity.

The Staffware environment gives the fulfilment centre an integrated picture of the total amount of work outstanding. It can prioritize and divide work effectively and efficiently. Workflow automation also supports the fulfilment centre in overseeing the whole process. An example of this is the monitoring of progress in other departments. Another is the ability to trigger outbound calls in the call-centre to follow up on offers that customers have not returned.

Integration with product systems means that more complex processes can be carried out by less experienced employees.

The future

Henk van Dijkhuizen summarizes the position: 'The first idea when we started the project was that the fulfilment centre would grow to need many employees. Automated management of business processes allows us to manage it with a relatively small staff. In the next few years, we will continue to add new processes. By encouraging customers to use the virtual channels, we expect a high growth in volume in the period ahead. The infrastructure needed to handle this is already in place.'

9 What to look for in a BPM product

In chapter 8, we mentioned the value of systems architectures. We deal with these at greater length below and with product matters. As elsewhere, we go far enough into the technicalities for you to be able to ask important questions of systems specialists and to place their answers in context.

We go no deeper, for two main reasons. The first is that any scrutiny of current products and technologies would consume space meant for – and possibly distract attention from – the broader aspects of BPM. The other is that it would get out of date quickly.

You will also not find in this book any recommendations for specific products or suppliers. Without knowing your situation, your objectives, your budget and other details, we could not safely do this. No conscientious consultant or systems integrator would make a 'blind' recommendation on such matters. No sensible software supplier would let you buy on that basis, either; it would be risking too much damage to its reputation.

We know how tempting it is, once you see an attractively clever product, to look around for applications for it. As an aid to getting the creative juices flowing, it has much in its favour as a starting point. As a basis for committing you and your organization to expense, disruption and hard work, there is nothing to be said for it. It becomes a solution looking for a problem.

As we discussed in chapter 8, when introducing new systems there is no substitute for a thorough, well-thought-out and collaborative programme of selection, design, implementation and management. We set out below some of the basics to take into account during the selection part of that process. If you wish to go beyond this level of detail, you should do so with the aid of experienced experts, from within or outside your organization.

The role of computer architectures

When selecting a significant piece of software such as a BPM product of any kind, you need to consider the effects on and from other corporate systems and standards.

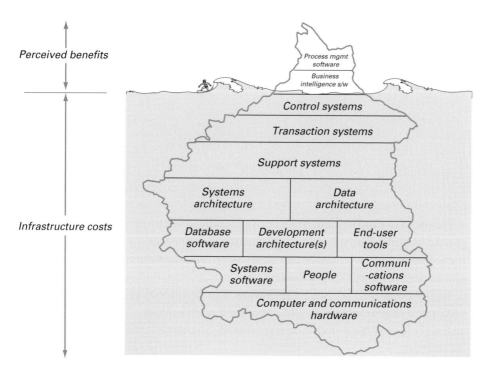

Figure 9.1 The systems complexity iceberg

This would be true even if you were setting up something completely afresh, on a green-field site, or deliberately going it alone. Sooner or later, you would need to import or export data in a common format, such as XML, or link to another process somewhere. (All the case studies show such interconnection to be the norm.)

The Workflow Management Coalition's interface standards, while a sound basis, are not enough to guarantee this on their own. The problem is wider and is a question of where BPM fits in your organization's overall systems. Figure 9.1 gives a visual (and unscientific) impression of the relative scale of the problem.

It is easier to visualize where BPM products fit overall if you place them in the context of an organization's computing architecture.

Dictionaries define architecture as the art and science of building. In its original sense, it applies to the use of stones, bricks and timber. In computing, it means a particular arrangement of combinations of hardware, networks and software.

Up to 1980 or thereabouts, anyone buying application software would not have concerned himself with the architecture of the product. It would have been pointless to do so. So long as the right hardware, network and type of terminal were present, in the right place and right quantity, there was nothing much else to discuss. It was a take it or leave it decision.

Application software then was usually made all of a piece, with no way to integrate other application software with it. Everything – user interface, programming tools, data management and directories – was knitted together by the supplier (more tightly in some products than in others, it has to be said). These monolithic offerings also often lacked the application programming interfaces that programmers today take for granted.

This was typical of the proprietary attitude of the time. The mainframe and minicomputer makers who dominated the market for this kind of software did not want other companies offering substitutes for their own products. Gradually that defensiveness changed, usually under customer pressure, and suppliers began to allow third-party products to work in with their own software.

It was the overwhelming popularity of the PC and its associated desktop programs that induced the real opening up of business software. Users liked the power and speed of response that they got from running programs on their own machine. They did not want to revert to using a 'dumb' terminal. Users soon demanded the ability to have their desktop programs work in with the corporate system.

From the late 1980s onwards, the major companies began introducing what they called client-server software architectures. In these, desktop machines were the client and the mainframe or mini was the server. These structures were intended to accommodate all kinds of application software. Manufacturing systems, logistics systems, payroll, CAD and applications development could all fit into the framework. Later still it became at least theoretically possible to use a wide range of third-party suppliers' products within these architectures.

The trend today is towards distributed computing, where the clients and the servers divide the work between them, sometimes interchangeably. Modern software architectures typically allow:

- the handling (and transmission and storage) of machine-readable data in all its forms – numbers, text, graphics, animation, image, video and voice
- a single system image: in other words, the system appears the same to the user no matter where or what he is accessing it from and wherever the process might be running
- the freedom to use a wide variety of access devices, from PCs (including Apple Macintoshes), terminals and workstations to portable digital assistants (PDAs) and cellphones
- integration with users' preferred desktop computer tools, such as word processors and spreadsheets
- unrestricted operation over local and distant networks, whatever methods and machines they might be based on
- remote working, over fixed or mobile networks, and unattended working (where the user's own machine can operate independently until reconnected)

- integration with other suppliers' software and systems, to allow the creation and management of enterprise-wide systems
- relative freedom of choice in the make, type and location of server machines.

As you may imagine, running a system containing all those ingredients is no easy task. This is one reason IT functions spend so much time, ingenuity and money in just keeping the whole software edifice in an operational state.

A general software architecture

Figure 9.2 shows a notional software architecture. It parcels into separate areas the way systems people look at how multiple computers work with each other over a network. Such a diagram is a conceptual model only and does not apply to any particular product or organization or any class of product. Much BPM software, for example, straddles more than one level.

Any layer can consist of several software products, which may be running sequentially or simultaneously. These may be hosted on a variety of hardware or operating environments Integration along a particular level can therefore be as much an issue as is integration between levels. We discuss those levels below, working upwards.

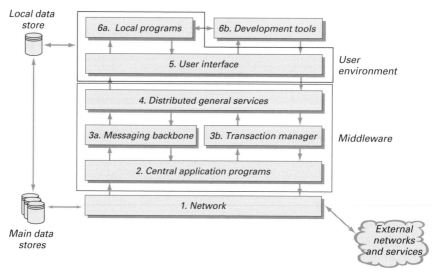

Figure 9.2 Generalized software architecture

Level 1: network

This level is fundamental to all corporate systems. Here is where you find network operating systems and protocols, such as TCP/IP and, decreasingly, Novell NetWare. It also includes wide-area products and methods, such as IBM's Systems Network Architecture (SNA) and X.25.

The cloud-shaped object at the bottom right represents external networks and services. By convention, we show this as linking to the network level. In reality, it could connect any almost any other level.

Level 2: central application programs

This is where the server element of client-server software usually is (remembering, as mentioned above, that that designation is mostly for the sake of convention). This layer also contains the server part of line-of-business and ERP programs.

Programs at this level typically work into a separate data or object store, shown on the lower left of the diagram. Normally this is a relational database. Process management programs use this to store rules, process designs and audit data.

Level 3a: messaging backbone

A messaging backbone is a communications hub or trunk between computer networks. It is commonly used way of linking multiple software products, for example, and is also important for inter-program and inter-process working. These subsystems work on the store-and-forward principle, on which data messages go into a software buffer for a short while before going to their destination. Many EAI products contain or work into a messaging backbone.

Level 3b: transaction manager

This element shares level 3 with the messaging backbone. It is shown separately because it is designed to manage real-time data traffic. This class of software is mostly used for traditional, high-volume data processing tasks.

Level 4: distributed general services

This layer includes all those general-purpose utilities and services that are required to mediate between servers and clients. Process engines, as shown in the centre of the WfMC model, belong here, as do separate rules engines.

These components, and the occupants of the two layers below (2, 3a and 3b), are middleware. If all computers used the same applications software, the same operating environments and the same databases, there would be no need for this. In reality, organizations doing distributed computing usually have to do so across varied systems. Middleware provides the tools and simplified programming interfaces to ease the difficulties involved in this. It also supplies supplementary services, such as security, systems management, gateways and indexing.

Level 5: user interface

There are three main graphical user interfaces in use today – Microsoft Windows, Apple Macintosh and Web browser (often applied within one of the other two). Although sometimes far from perfect, these have at least imposed a degree of commonality and consistency among applications programs for personal computers and other devices for users.

Level 6a: local programs

These are the application programs that run on what is often called 'the desktop' (even though it could be nowhere near a desk). These programs include the traditional personal productivity tools of word processing, spreadsheet, diary, and email client. Programs at this level normally include a local data store, shown at the upper left. This links (or should do) to the central stores, each reciprocally updating the other.

Level 6b: development tools

Here is where programmers' and designer's tools notionally fit. These include process modelling and simulation programs. Not every user would have or need these.

Using a framework like this allows you to describe and analyse the systems in any organization that uses computers extensively. You can apply across a supply chain, as well. Figure 9.3 gives an imaginary example.

Your organization is at the top left. The programs it uses link to the relevant programs in your trading partners' systems, through the networking cloud. Supply chain integration and end-to-end process management demand that this be possible and that it work efficiently. It does not, however, demand that every partner's architecture be the same or even that the programs they use be the same. Partner organizations do not even have to disclose their architectures to each other, although they typically do. It aids problem solving when architecture speaks to architecture.

Thinking architecturally

One of the main benefits of setting out such an architecture, even though it may not accord with any physical or operational reality, is that it aids thinking. It is a model whose deficiencies people willingly accept in return for the ease of reference and the common language it gives. Just saying 'at level 3' or pointing to it, for instance, immediately allows other people to place in context what you are saying. It also helps impose discipline if someone proposes the adoption of a new service or piece of software. Some organizations make it a rule that when doing so, that person must say where the proposed addition sits in the architecture and how it will interwork with existing entities.

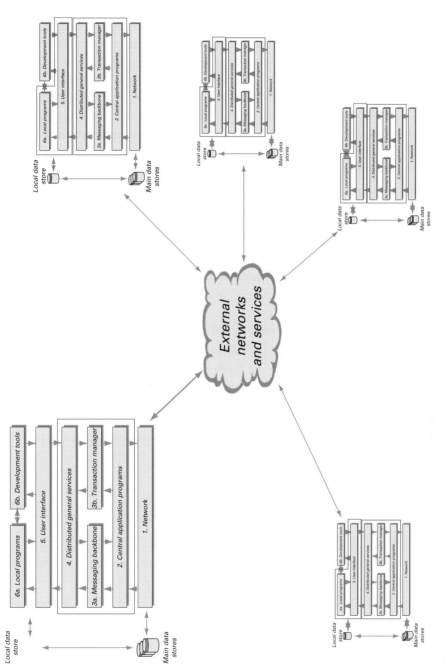

Figure 9.3 Linking with trading partners

We have presented a generalized, all-purpose architecture in the diagrams above. If your own organization has or is creating its own systems architecture, it will almost certainly differ from this. It must be individual to your organization's situation, needs and aims. Slavishly copying someone else's architecture or adopting one 'off the peg' from a supplier or consultant is risky. It may seem a quick and simple way forward at first but it would be unwise. Your systems people would soon be spending so long trying to bend it to the organization's needs that its deficiencies would impede progress.

Bear in mind also that an architecture exists separately from the objects and services that make it up. It is, metaphorically, above and beyond its components. Indeed, one of its purposes is to make it easy to safely remove any of those components and replace them with something better or cheaper. The architecture should outlive its components and withstand their replacement. It should also be able to satisfy the needs of the organization, however its structure or functioning may change.

Given the intricacy of the typical large company's architecture, its systems department normally used a computer to draw it up. This allows the systems staff to capture the full detail of specifications and components. The task is eased if this detail is held on an underlying database. As with process design tools, this gives the necessary adaptability while keeping track of components and how they change. The similarity does not end there.

Creating and maintaining a systems architecture is a continuing process. It needs to be managed and, as with any design process, is best done collaboratively with users. Nobody can hope to draw up from scratch a detailed architecture for the whole organization. It makes more sense to start small and locally. You can then grow and rework that micro model into the macro model.

A danger to avoid here is ending up with a diverse and incompatible set of small architectures, which is why communication and collaboration throughout are vital. The 'balkanization' of computing architectures would be worse than having none. Whatever the scope of your particular architecture, build in to it the minimum necessary to meet the needs of today, while leaving the maximum freedom of choice for tomorrow.

Choosing a product

A computer product of any kind does not just sit within the context of the systems and architecture into which it is intended to go. Much like a tree or the Marble Arch (see chapter 1), it also represents a point in several other processes and contexts. It has a development history (and, you hope, a future), its maker has a history (likewise) and, if you buy through an intermediary, so does its supplier. These histories and contexts make a difference to the way a product is constructed, supported and sold. Organizations do not forget their histories, either; they are embedded in their traditions, their structure and their ways of dealing with the world.

These extrinsic matters are easy to overlook if you concentrate just on the abilities and features of a piece of software. When looking at product, therefore, it helps to think of as coming wrapped in two concentric layers.

Immediately outside the product itself is the offer of which it is part. This consists of such matters as contractual terms, and the support and training offered. It also includes help with 'porting' the product. (Porting is the complicated task of making software able to run on a different operating environment, such as moving it from MS Windows to Linux.) These and other similar factors need evaluating in addition to the product's features.

Outside the offer layer comes the supplier layer. This gives the context for the offer and includes such considerations as the supplier's financial, organizational and human resources, its alliances and its trading history with your organization (if any).

Here are three short lists, each in alphabetical order, of some of the matters to assess in those three layers.

Product aspects

- Its architecture (and does it mesh with yours?)
- Customer experience
- Ease of migration
- Future proofing
- Hardware integration and compatibility
- Installed base
- Management and administration tools
- Networking options
- Performance
- Quality and reliability
- Software integration and compatibility
- Special features
- Standards compliance
- Training needed
- Upgrade paths
- User interface.

Offer aspects

- Availability
- Customer experience
- Finance

- Licensing scheme
- Porting help, if necessary
- Price and terms
- Product futures
- Project help
- Servicing and maintenance
- Support
- Third-party products and services
- Training offered.

Supplier aspects

- Does the company have a customer focus or does it concentrate on technical matters?
- What is its image in the market?
- What are its trading and technical alliances?
- Will its culture match with your organization's?
- What is its corporate history? Is it turbulent or stable?
- Does it have a trading history with your organization? Is it satisfactory?
- Does it have enough channels to market? Are they in all the countries and areas your organization operates from?
- How do you rate its financial, organizational and human resources?
- Can you believe what its employees say? Do they and it keep their promises?

We are preaching perfection here but your organization should buy only when it has satisfied itself that all three levels match its needs. A riddle wrapped in a mystery inside an enigma is of no use to it. Where there are competing suppliers, or competing methods of acquiring a system, it will have to score and rank each of these factors. (At the risk of labouring the point, this too is best done with the active involvement of interested parties.)

What a BPM product should offer

We now look specifically at products for Business Process Management. In table 3.2, Process management software compared, we set out the differences between the capabilities of a full-blown BPM product and those of other popular types of process management software. In table 9.1, we have stripped out the middle columns of that table to concentrate on BPM software.

You can use this list as the basis of a set of product selection criteria. It is unlikely, though, that you would find many or perhaps any products that could carry out all the listed tasks. You would probably have to consider using

Table 9.1 Target BPM product capabilities

Ability		BPM
Data handling	Makes electronic versions of paper documents?	Typically uses an EDM system for this
	Handles all kinds of digital document (e.g. word processor files)?	Yes
	Indexes and searches digital documents?	Uses EDM for this
	Allows the reuse of documents, and the disassembly and recombination of their components?	Uses EDM for this
	Allows users exclusive access to master documents (check in/out)?	Uses EDM for this
	Links digital documents with other data?	Yes
	Exchanges data with other software?	Yes
Routing	Routes digital documents around a network?	Yes
	Manages the speed and direction of the route?	Yes
	Sets up rules for managing routes, timing, actors, etc.?	Yes
	Runs processes outside the organization?	Yes
Process management	Modifies processes on the fly?	Yes
	Manages human activities?	Yes
	Invokes the running of other software	Also integrates and interacts with other software
	Manages an organization's other software (orchestration)?	Yes
	Uses, and is used by, Web services?	Yes
	Offers goal-directed processing?	Sometimes
	Offers predictive processing?	Sometimes
Process design	Graphically models processes and creates the rules to run them?	Yes
	Simulates and optimizes processes?	Yes
	Manages the rules from application programs separately?	Yes
	Allows analysts and users to add or modify rules?	Yes
Reporting	Provides after-the-fact reports and audit trails?	Yes
	Provides real-time reports on progress and performance?	Yes. Also provides BAM
Product design	Open programming interfaces?	Yes
	Standards based?	Yes, always

combinations of products, perhaps from different suppliers. This would especially be the case if you had software already working that does some of these tasks adequately. A further consideration would be the kind and make of products that your supply chain partners might be using. (We discussed this in chapter 5.)

Here are some further aspects that you or your technical advisors might consider, under similar headings to those above:

Routing

The software should be able to work across every kind of network you have in your organization and those that link to your customers and trading partners. It should offer both conditional and parallel routing.

Process management

The software should be able to manage processes from end to end, irrespective of the type and number of other processes, application programs and machines involved. It should therefore possess extensive and extensible integration abilities, working with internal and external application programs. Also, it should have an orchestration tool to enable processes to call and bind to external business processes when needed.

Process design

There should be separate, graphically based tools for modelling (preferably with simulation) and for process design. Both should be easy to use and understand. They should be able to record both manual (human) steps and automated steps. There should also be a forms designer for human activities.

The process designer or definer should allow system designers to optimize processes and debug them. It should offer a sufficiently a wide ('rich') range of basic process elements. Designers should be able to reuse procedures and sub-routines. There should also be a good set of process templates or patterns available, suitable for use 'out-of-the-box' if you wish. Look for modifiable process frameworks designed for your industry, such as for mortgage processing, claims management or telephone order management.

Be sure that the design tool can create processes that deliver work to groups as well as individuals. It should also allow for changing work roles. Processes should be able to deal with peaks and troughs, holidays and sickness, work redirection and unplanned business events. Deadlines and escalation procedures should be definable. It should also be possible to create 'skinny' processes, in which the developer does not need to know the full complexity of a process, or all the options available within it. These processes are instead built on the fly, adapting to the specific case.

Reporting

Work queue management is a basic requirement, as is the reporting of progress on individual work items. Modern products should offer a degree of prediction of, for example, the start and end dates of work items and the workloads of groups and individual users. *Ad hoc* exceptions to defined procedures should be possible. There should be full reporting and an audit trail for all actions and data movements.

Product design

The choice of which process management engine is right for your needs will depend on the volume of transactions to be processed and the degree of human intervention required. You may have to decide between, in suppliers' parlance, best-of-breed products and those offering more moderate abilities across a wider range of situations.

Whichever product or system you choose should offer high performance, being able to handle many processes, users and links to other programs. It should be 'scalable' – that is, able to expand without difficulty from a small system to a large one. Reliable round-the-clock running is a necessity, as is good security of data and designs. Look for failsafe and 'fail soft' working, including 'roll back' in case of networking problems.

The software should be able to run on single or multiple servers, in whatever arrangement these may be in, across a wide range of operating systems. Also, it should work into the main types and makes of database, keeping process designs and data in a single repository. Ease of use, installation and maintenance are important qualities, as is the ability to provide human interfaces in the main languages.

Choosing a supplier

Whether introducing BPM software for strategic or tactical purposes, you need to evaluate the supplier or suppliers. These could be the software maker, a system integrator, a consulting firm or any of those three.

To begin with, the prospective customer must ask himself if he thinks that the software maker will still be in business, in a recognizably distinct form, in five years or more. Get the help of industry analysts or a stockbroker when making this assessment.

If the answer is negative, or even qualified, then you need to consider the possible consequences. It could be that you feel confident that better standards will deal with this. Alternatively, you may feel that any purchaser of the company will continue to support and develop that product long enough for your purposes.

It is common when assessing potentially important suppliers for your technical specialists to meet with those of any potential supplier. They can then readily assess the supplier's technical strategy and to see if it is congruent with your own. You should think twice about continuing to deal with any supplier unwilling to share this information.

You should also consider the your future relationship with suppliers. They find it hard to resist the temptation to 'lock in' their customers, either by technical means, with proprietary standards, or through contractual arrangements. Some user organizations are happy for this to be the case; others are not. Either way is acceptable so long as you choose it with your eyes open.

Supplier relationships can be categorized as falling mainly into one of four types:

- The simplest is where the supplier is concerned only to ship relatively uncomplicated goods or services, at competitive prices. This is the *commodity supplier*.
- Next is where the supplier offers to solve a particular but bounded business problem, typically with a combination of product and service. These organizations call themselves *solutions suppliers*.
- Then come the *systems integrators*, in which the service element forms a major, and sometimes the entire, offer. Often, these companies are asked to integrate the various 'solutions' that have been introduced earlier, to make them accessible to a wider range of users and other systems. Systems integrators frequently come in to help create enterprise-wide systems.
- The final entity is the *information utility*. This is almost an anti-product offering. These organizations are typically large computer, consulting or telecoms suppliers who can take over responsibility for the entire computing (and telecoms) needs of an organization.

Each of these supplier roles imposes different requirements on the customer, creates a differing level of dependence on the supplier and has a different cost structure. Once past a certain size, any supplier can occupy all or any of these roles. IBM, Unisys and Hewlett-Packard do, for example. Conversely, any customer can occupy any and every role as a purchaser if it so chooses.

Make, buy or rent

The decision among these three options is standard for all kinds of application software. You might want to consider all three for BPM, although, in our view, your organization is unlikely to want to write its own software. The whole area is too new, too complex and too fast-moving to make that worthwhile for all but the technically mostly confident and competent.

The decision then falls between buying and renting. Software suppliers are mainly set up for the former. A small but increasing number is looking instead to offer access to BPM software on a metered basis or by agreed time period, such as monthly. This would be an attractive option if, say, your organization or your part of it had few computing staff of its own. It also reduces the size of the initial investment and exposure to the risk of project failure.

A near cousin to this approach is BPO, in which the user organization farms out the management of its internal processes. Typically, this sort of service comes from large systems integrators and computer companies.[1] The information utility idea mentioned above is BPO taken to an extreme.

The attractions of outsourcing processes include potential savings in capital and operating costs, less demand for internal systems expertise and releasing managers' time. The other side of the coin appears in users' reports of inflexible contracts, difficulty in managing suppliers and a lack of measures of performance. Also, not all processes are suitable for handling by somebody else. The most apt candidates are usually those that are repetitive and unvarying and that contain a great deal of human work.

A future potential difficulty comes with the growing influence of corporate governance legislation, such as the Sarbanes–Oxley Act in the USA. This can require a company director or executive to be personally answerable for the correct completion of certain processes. Self-preservation will disincline that person to hand such responsibility to an outside body. Time will show how service suppliers will provide the reassurance these users will want.

When assessing potential suppliers of process services, you can use the product checklist above to choose among them. You should not be expected to put up with inferior processes or process management simply because you do not control either directly.

Futures

BPM is still a young subject, with new products and technologies accompanying it.

As with any new computing market, there are dozens of suppliers of pre-existing products who are rebadging them as BPM software. Sometimes this is reasonable, with the basic abilities of the product matching most of the criteria we have set out above. At other times it is done on the basis that there is one born every minute. The stance of these suppliers is that if they say it loudly and often enough, people will believe their product actually is suitable for BPM. This is cheaper for them than building something new or reengineering the existing product line. It is the sort of thing that happens whenever a new market emerges; you need to be on your guard against it.

---

1 The word, 'outsourcing', is quite ugly enough for some people's taste. When a company in a different country provides this service, it is known by the even uglier term of 'offshoring'.

Genuine new products conform to some changed ways of thinking about how to make and use process management software. There are three major changes confronting users and designers of BPM systems:

- the rise of service-based approaches to computing
- how systems provide business activity monitoring
- how organizations buy and use process management tools.

We discuss each below.

BPM in a service-oriented world

The way organizations manage their business processes continues to evolve. Formerly, they had (and still have) workflow and EDM. Today's solution – described in earlier chapters – is to combine the basics of these with system integration tools and activity monitoring to give BPM software. Process management systems are now entering a third stage, which is about managing computing services and orchestrating them.

Computing services allow programs or entire systems to call for or provide services to other programs and processes. To take a simple example, the graphics part of a word processing program might send a command to a printer or retrieve a piece of clipart from storage.

A resource could be on the same machine, on the same network or separated by the Internet. Working like this over the Internet is often called Web services; this uses XML and other standards to define its components. The rise of middleware and the emergence of the Internet as a computing infrastructure have made this move to services possible.

Whether communicating through Web services or some other method, service requesters and providers should each be indifferent to where the other is. The requester simply sends out the call and, provided it is valid, the provider responds to it.

This kind of arrangement is these days known as 'loosely coupled computing'. The older style of working is naturally now called 'tightly coupled computing'. In this, a program always communicates with the same, known resources.

There are, of course, many ways of defining how different programs and applets communicate in a loosely coupled arrangement or architecture. Equally inevitably, there is hot debate among their adherents about which method is the better. These debates do not concern us here. They are conducted at a high technical (and emotional) level and change almost weekly.

Corporate computing is thus increasingly a matter of connecting various computing service providers with requesters. This is changing the way organizations deliver businesses services to their customers and interact with their suppliers, partners and employees.

Orchestrating these programs and services has become inevitably more complex. Higher levels of mass customization and personalization at Web sites demand a widening range of choices and interactions. Also, the technical environment is no longer the closed and controlled system long familiar to IT departments. It is open, unpredictable and rich with exceptions.

The result of these innovations is that organizations using these loosely coupled arrangements are no longer in direct control of the services other organizations provide. Failures in delivery or processing, incomplete information and unanticipated responses are now the norm rather than the exception. Efforts to develop high-value application programs in this new business environment have proved mainly to be:

- costly
- hard to change and extend
- hard to validate
- susceptible to failures and exceptions
- unable to grow easily ('scale')
- unable to deliver the expected value.

From a technical perspective, workflow and BPM both provide users with a process architecture that has well-defined initiators and participants. Processes can be 'workflow enabled' and made up of reusable sub-processes. Organizations can encapsulate the underlying process logic in a readily understandable way, which helps them become responsive ('agile') and adaptable ('flexible').

The newer approach, of orchestrating and managing services rather than programs, involves creating something called a service-oriented architecture (SOA). This separates the what (processes and metrics) from the how (resources provided by events and managed services). In short, processes become managed services. These can be loosely coupled, component-based and usable on various kinds of computer and operating system. Above all, they become reusable.

Organizations need something to coordinate the interworking of all these reusable process elements. We call it the process event manager (PEM). This will take charge of and synchronize the way process components are extracted and linked to meet a specific process need. It will do so according to user-defined process templates that reside within the BPM software.

Trends in BAM

Organizations are under constant pressure to become more responsive, improve customer satisfaction, comply with regulatory requirements and shorten time-to-market. All these impose the need for up-to-the-minute and comprehensive management information, giving the so-called 360-degree view of the company. BAM is the usual means to gain this view. Often, as we have discussed in earlier chapters, it is part of a BPM system.

Earlier business intelligence tools provided only simple reports and often not in real time. Modern BAM tools work with the BPM software to give near-instantaneous feedback, combined with process orchestration, sophisticated event handling and *ad hoc* process management. Ideally, the combined system achieves these results without needing to develop intrusive computer code that will affect existing application programs.

Whether and how this is possible will depend on circumstances, which fall broadly into three camps:

- when the process is contained in several existing application programs and cannot be extracted from them without an expensive and slow rewriting of those programs
- when the process can be easily defined, engineered and implemented as a BPM system
- a combination of the first two, where neither approach will fully meet the organization's needs.

In the first of these, the internal systems occupy their own domains (silos, in the jargon). We have argued earlier that these systems would be better served if an independent process layer controlled them. This is not always feasible. A hybrid answer is possible instead. The BAM tool is set up to monitor and manage the interaction of these systems, dealing only with exceptions. It can then pass the processing of these exceptions to the BPM software, which can handle them quickly and suitably.

The second situation is where BPM comes into its own, permitting fast development and implementation of systems. Here, users have recognized the need to reengineer their systems and take a more process-centric approach to their introduction and use. This is 'sweet spot' BPM software, where a process suite fits best. Such a system would give genuine real-time monitoring and management rather than the near real-time response that reporting tools offer. The advantages of this include:

- automated solutions and dynamic re-routing of work
- easier integration into systems management systems, such as Tivoli
- monitoring of sub-flows, such as those triggered within process orchestration and Web services.

The third, mixed, choice combines the other two approaches. There are often situations where only parts of an organization's systems can be reengineered (option 2). The rest consist of 'silo' programs that cannot be touched (option 1) but which need to be part of an overall BPM strategy.

A typical example is the way telecoms providers manage complex orders. An important part of this is known as the provisioning process.[2] There are many separate back-office systems that form an important part of telecoms provisioning.

---

2 'Provisioning' is telecoms-speak for providing. It normally implies supplying a user with everything that he, she or it needs to use a service, such as equipment, wiring and transmission capacity.

Often their complexity prevents them from being integrated into the main process, yet they run 'micro' or sub-processes that need to be monitored. If a delay occurs in one of these background systems, the effect on the overall process can be significant. It is important, therefore, to monitor and manage the interactions between the micro processes and the main process. Few if any BPM products can do this today, yet the solution is relatively simple.

If business activity monitoring is to meet its promises it must be able to accommodate all the demands and possibilities set out above.

On-demand process management

Many people see BPM as the preserve of large organizations. The cost of the software and the difficulty of integrating it with existing programs can make it demanding of financial and technical resources.

Another common problem, which this book should help you avoid, is a mismatch of perceptions of BPM and its role. Computer departments sometimes view BPM software only as part of the technical infrastructure. They regard it solely or mainly as a means for integrating systems and developing new applications. This, as you know, is far from the whole story. Users, by contrast, see BPM as an aid to reducing errors, completing tasks sooner and better and easing the task of running an organization.

One way to minimize the drain on resources and to ensure that the emphases are in the right place is to use BPM as a service. Suppliers are starting to target the BPM market with what they call process-on-demand, a form of outsourcing. In this, the supplier runs the process software on its own systems. It links to the user organization's system only when needed and typically charges according to capacity used. The supplier's motto is: 'Use what you need when you need it, and no more.'

The benefits of using a BPM service are much the same as those for any farmed-out service. Initial and continual investments are modest, disruption is lessened, speed of implementation is greater and demands on technical staff are slighter. Against it, also as with any bought-in service, are the loss of direct control, the risk of inflexibility in supply and the potential of lagging behind technically. These must be balanced against the advantages but, overall, this is a solution worth looking into, especially for small-to-mid-sized organizations.

There will doubtless be other trends arising and suppliers coming to market with new ideas. You can stay abreast of these by using some of the sources of information we list in the suggested reading section at the end of the book.

Case Study 8 Sonae Distribuição Brasil

Sonae Distribuição Brasil (SDB) is a large retail group in Brazil, with supermarkets and other stores in sixty-two cities. Margins are tight in this sector, especially in Brazil, and there was evident waste and inefficiency in some internal processes. Administration and expenses accounting were primary targets for improvement.

The group called in a systems integrator specializing in workflow and experienced in the database software SDB uses. With extensive user involvement, this firm created a bespoke process management system. The first application of this was for travel management. Two years later there are fifteen systems, covering sales management as well as administration.

Financial savings of over US$2 million have resulted. The internal processes to which the software has been applied are faster and less open to possible abuse. Although only indirect beneficiaries, customers report better and faster dealings with SDB. The group itself feels better placed to cope with change.

The SDB installation has the most users of all the eight cases we show in full. Despite that, the company and its systems experts placed great importance on users' involvement in the design process. Perhaps, as with the Matáv case but in a different direction, this is typical of the sector, the country, the situation or some combination of the three. Generalizing about such matters can be risky.

Industry/ Sector	Retail stores	**Location(s)**	Brazil
Annual turnover/ income	BRL3.4 billion (US$1.2 billion)	**Number of employees**	21,000
Type of system	Workflow	**Supplier and product**	Oracle Workflow
Number of users	Up to 3,200, depending on process	**Time to complete**	2 years, for 15 systems
Business objectives	Saving costs, speeding decision-making, improving customer care and standardizing best practices by integrating the company's activities		
Quantitative results	<ul><li>Average time to approve expense requisitions fallen from 10 to 2 minutes</li><li>US$2 million a year saved though failure elimination</li><li>20,000 hour annual saving through task automation</li><li>70 per cent of invoices now processed in the month of issue (40 per cent previously); US$33,333 annual saving</li><li>1,500 hour annual saving by eliminating progress-chasing phone calls</li><li>Total benefits of over US$2 million</li></ul>		

Qualitative results The most important advantages are:
- the agility of processes
- standardization of best practices
- the certainty that processes will work as designed

Business background

SDB is the third largest retail group in Brazil. It has 166 stores in sixty-two cities in four southern states of the country, with its headquarters in Porto Alegre. In 2002, its sales reached 3.4 billion Brazilian Reals (BRL). SDB is the tenth largest employer in Brazil, with 21,000 employees.

The company is part of the Grupo Sonae (Sonae Group), the biggest non-financial private group in Portugal. This is active in many areas, such as retail, lumber, real estate, telecoms and venture capital. Grupo Sonae operates in more than fifteen countries in four continents.

The retail industry in Brazil is highly competitive. It is based on large volume and low profit rates. The market is going through a rapid consolidation, with the market share of the four largest companies growing from 27 per cent to 43 per cent in the five years to 2003. This points to even tougher competition in the future. Keeping and increasing market share and leadership will be hard.

Since its arrival in Brazil, SDB has aimed for low prices as its main competitive advantage. This is explicit in some of its slogans, such as 'The champion of low prices!' and 'You know BIG is cheaper!' (The latter is a pun on the name of one of SDB's networks.)

The company's market positioning was being threatened by the difficulties SDB had in managing the processes in its stores. The flow of paper around its widely scattered premises formed an obstacle to productivity, efficiency and control. A process could take days or even weeks moving from store to store. Corporate headquarters would not even know about its existence. This resulted in a collective sensation of lack of control, many times nearing chaos.

These problems were compounded by the poor economic situation in Brazil. Companies faced stagnation or even a fall in sales. SDB was no exception, and it thus became essential for the company to reduce its costs. Administrative expenses in the stores and headquarters were a prime target. SDB felt that workflow automation would make it possible to balance these. The approval process would be better organized, as well as safer and more trustworthy. SDB also saw in workflow technology a tool to integrate the company and to make geographic and organizational limits irrelevant to performing any process.

In 2003, workflow automation was extended to retail-specific processes, such as product advertisement approval, customer claims and sales campaigns. These processes lacked effective control and often were not properly followed.

System description

SDB chose Oracle Workflow as its workflow development and execution software. This runs on an IBM RS/6000 Enterprise Server running AIX. Other software mounted on this includes Oracle Database Server, Oracle HTTP Server and Microsoft Exchange Server. Workstations accessed these through Microsoft Internet Explorer and Outlook. Other, existing systems were SAP R/3 and the servers for three bespoke systems:

- BDN (Base de Dados de Negócios – business database), which holds details of suppliers, department and stores
- RH (Recursos Humanos – human resources), holding information about hiring and firing employees
- SGO (Sistema de Gestão Orçamentária – budget management system), containing budget details of every account.

There were fifteen workflow systems in use by the end of January 2004. They manage twenty-five business processes, formed by 174 interactive activities and 198 automated activities. These range from simple relational database queries, with routing, to complex integration with existing systems.

SDB's workflow systems fall into two groups:

- General and administrative processes deal with travel, on-credit expenses, reimbursement, cash expenses, payment requirement, IT project control and investment.
- Retail-specific workflows include goods price change, advertisement approval, client claim, sales campaign, client feedback, product divergence, supplies budget change and supplier information change.

The cells in figure C8.1 with negative results show that the cost of this journey exceeds the available budget. Budget data comes from the SGO system and user data from the RH system. The icons at the bottom allow the user to view the attached documents.

One of the newest features in SDB's workflow systems is the use of Wireless Application Protocol (WAP) to transfer process results to cellphones. The telephone's screen displays only essential information – requester name, supplier, available budget, size of expense and new available budget. In the last two lines are the routes available to the user – approve, refuse or forward to another manager. When the user selects one of these, the WAP software communicates online with Oracle Workflow Server and then ends the task.

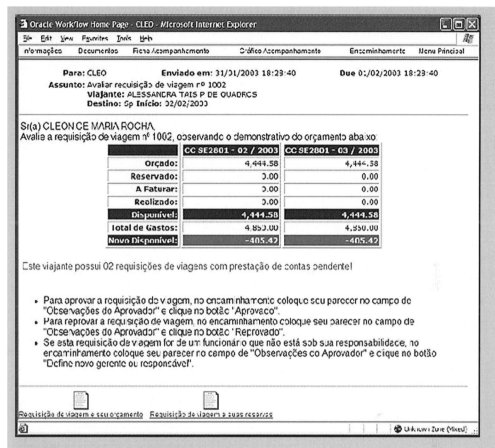

Figure C8.1 Travel approval notification

Implementation experience

Where and when did the project or system originate?	Development of the first workflow system started in January 2002, after a successful pilot project
How long did implementation take?	4 months for the first workflow system, for travel management; this began development in January 2002 and went into production by the following April
Who did the implementing (own staff, contractors, consultants, etc.)?	iProcess Soluções em Tecnologia, an Oracle partner specialized in workflow and content management, developed all the systems
How much bespoke development was there?	All the systems involved bespoke development; this included application creation (such as travel requisition forms), systems integration and the workflow process itself
Were there any special infrastructure needs?	None; Oracle Workflow runs inside SDB's existing Oracle database server

What were the most significant implementation issues and how were they dealt with?	1. Integration with existing systems, especially SAP R/3; these arose from the team's lack of experience in using R/3's integration resources 2. Workflow system performance, which suffered when the workflow database grew fast; cured by redesigning earlier workflow applications and designing new systems with these issues in mind
Who is responsible for the system overall?	SDB's IT department is responsible for all systems
How was and is training handled?	SDB created education teams of users from all the departments involved in each system, each team underwent an intensive training from the development; afterwards, each education team wrote an operational manual for users An employee in every department is the appointed system expert; this person has learned the system and helps workmates with immediate problems
What was and is done to encourage use?	• Deep and constant involvement of managers and leading users • Allowing users to drive design
What lessons were learned?	• Gain and use political support • Involve users throughout all phases • Pay close attention to usability

Developing the first workflow system began in January 2002, after a successful pilot project. SDB's senior managers were searching for a process management system and found that workflow automation was the right tool to promote the changes they knew SDB needed. The system went into production in the following April. The results achieved stimulated the development of new systems, which now total twelve in number (see *System description*, above).

A technology such as workflow, which promotes drastic changes in the company, can be implemented only with significant political support. Fortunately, the leading users and senior managers were enthusiastic for it. The pilot project, which took only three weeks to develop, convinced them that workflow software was the right tool to promote the changes they knew that SDB needed.

SDB's managers were in a hurry to put the systems to use. The first system was introduced at a frantic pace. Anxious to achieve the desired results, senior managers held back no resources. They demanded that systems be developed in weeks, rather than the more usual months. Even the CIO got involved in the interviews and definitions of the first phases of the project.

Approach

To speed development, the project team adopted a JAD approach. Instead of having individual meetings with every department, large meetings were scheduled with all the participants together. These included the most experienced people and those who really knew how the work has done. Their managers were also invited, so that decisions could be made during the meeting. The unconditional support given by senior managers made those intermediate managers feel comfortable about redesigning their processes.

User involvement was seen as crucial. In all the stages of the development, the final users were present and made decisions about all the aspects of each system. Users were encouraged to behave as leaders of technological innovation: they never had to accept it as something finished. Leading users had the authority to change processes. One result was that the IT department often needed instead to worry about how to handle the huge number of demands. There was no question of their having to 'sell' the system to the users.

Redesigning this number of business processes normally takes weeks or months to complete. In SDB, a workflow analyst, working with leading users, was generally able to do so in days. Changes were speedy because they were at the operational level, being neither tactical nor strategic. Most workflows focused on doing the same things in a more efficient way, rather than deeply reviewing processes. One good example was the automation of invoice entry in SAP, which saved thousands of work-hours. Another was the elimination of paper flow.

Another reason for speed was that the workflow analyst was free to suggest ways to improve processes. There was no process standardization department to satisfy first.

SDB felt that investment in usability was essential. In their view, the standard tools in workflow products would not give good enough user interfaces and created bespoke workflow applications instead. This increased cost but the company felt the return has been worth it.

Training

User training needed a huge mobilization. There were hundreds of departments and thousands of employees to be trained. They had to become skilled in using the systems. An inability to run a critical process, such as a price change, could make their department stop.

SDB created education teams of users from the departments involved in each system. These teams underwent an intensive training from the developers. Afterwards, each team had to write an operating manual for their users.

In every department, an employee was chosen to be the system expert. This person learned the system and helped his workmates. The education teams watched the progress of these local experts and kept them up to date with what was happening and when. By keeping users continuously informed, there were fewer post-release problems than expected, despite the fast system implementation.

A potentially tricky problem was the dissatisfaction of a few employees who had previously taken advantage of the lack of control to act dishonestly or unethically. Workflow systems put so much light on the processes that there was no way for them to continue their damaging practices. These people could not openly reject the systems and therefore had to accept the new rules.

Technical matters

The main technical hurdle was in integrating the workflow software with the existing systems, especially the SAP R/3. These problems arose mainly from the team's lack of experience in using R/3's integration resources. As a result, some systems were first released with manual integration, being fully integrated some weeks later.

Performance was another issue. When the workflow database started growing fast, performance started to suffer. To cure this, some of the earlier applications were changed. New systems are developed to avoid the problem. Today, the database holds around 600,000 workflow instances (runnings of processes).

Workflow technology is now seen as the missing link in connecting a wide range of information systems in single, streamlined processes.

Benefits and user reaction

What has been the reaction of managers and staff?	• Enthusiasm from managers and leading users • The ease and interactivity of the new system helped win users over • Initial fears of job losses were unfounded The system is not used to reduce jobs
What has been the reaction of customers or trading partners?	They do not directly encounter most workflow systems but do experience the benefits in improved quality and response time of SDB's processes
What has been the overall cost of the system?	Not available
What have been the main process benefits?	• Greater control of costs • Speedier processes, some reduced by more than half • Better-informed decision-making • Increased process awareness • Standardization of best practices • Eliminating failures • Freedom from constraints on workplace location • Better support of customers.
What have been the main effects on operating style and methods?	Has allowed SDB to define best practices and standardize them in every of its hundreds of stores and departments; the level of automation achieved brought much more agility to the processes – always a need in retail

Most leading users and senior managers received the new system with enthusiasm. They were sure that workflow automation was the right tool to promote the changes they knew that SDB needed.

The simplicity and productivity of the screens designed won most users over. Its ease and interactivity sharply contrasted with existing systems. When users realized workflow was helping them to work faster and better, they immediately started supporting it. SDB is a time-oriented company and making better use of time is important to all employees.

Some employees, mainly secretaries, were groundlessly afraid of becoming useless and being fired. In any event, none of the systems was promoted as a staff reduction tool.

Customers and trading partners do not directly encounter most workflow systems. They can, though, see the improved quality and response time of SDB's major processes, especially through its the supplier portal.

Introducing workflow technology has transformed SDB in many ways. Most stem from the new-found integration of the company's activities. This has made it possible to marry operational decentralization – an unquestionable need of the retail area – with operations control – a central interest of senior managers.

SDB, despite being geographically dispersed, has now started to perform in a more coordinated and efficient manner. This is showing itself in several ways:

- greater control of costs
- speedier processes, some reduced by more than half their previous cycle time
- better-informed decision-making
- increased process awareness
- standardization of best practices
- eliminating failures
- freedom from constraints on workplace location
- better support of customers.

An example is in the price change mechanism. Each store manager needs to be free to set the prices of goods, based on local factors such as a sale in a nearby competitor's store. Internal auditing detected anomalous behaviours in some stores. Products were sometimes being sold at a price that was a tenth or even less of the original cost, causing a loss to SDB. Using the workflow system, proposed price changes must now first be approved by the respective manager at headquarters.

Although not a major objective to start with, the higher speed of processes has become a highly valued aspect for managers. The retail sector is one of constant urgency, where decisions are taken fast and deadlines are strict. Workflow technology matches this culture perfectly.

Extending workflow automation to retail-specific processes in 2003 improved customer care and helped eliminate process failures.

Quantifiable benefits

These include:

- A saving of 75,000 workhours a year through task automation. These stem from automating complex routing rules (mainly based on relational database queries), electronic notifications, automatic update of databases, automatic control of deadlines and automatic data input in SAP R/3 and other systems.
- A fall in the average time to approve a requisition (expense, travel, etc.) from 10 to 2 minutes. Roughly 160,000 requisitions were approved in 2003 and every requisition involves an average of 2.5 approvals. This equates to a 53,000 hour saving in managers' time.
- A halving of the number of staff inputting data to SAP R/3. Most of the data is now entered automatically by workflow systems.
- The elimination of failure in the price change process. This saves an estimated BRL6 million annually (US$2 million).
- Faster processing of invoices. Before workflow, only 40 per cent of invoices were processed in the month they were issued. This led to fines for paying taxes late. Now, SDB processes 70 per cent of invoices on time. This produces a yearly saving of BRL100,000 (US$33,333).
- The virtual elimination of telephone calls asking for process status. A self-service workflow monitoring application removes the need for these, saving around 2,500 hours a year.

Although impressive, these numbers do not reflect the most important advantage to SDB. This has been the discipline brought to every decision taken, in every department. Everyone now has a clear view of the results of his acts. The workflow systems have made employees partners in the search for more efficiency.

The future

SDB sees workflow technology not as a system or application but as a tool to become increasingly efficient. Expanding its use is something strategic. Senior and IT managers decide which business processes should be changed and new systems are developed fast. This increases the trust in the technology and stimulates managers to ask for more systems, building a 'virtuous cycle'.

Sonae Distribuição, the Portuguese company that owns SDB, is now excited by the possibilities of workflow automation. It intends to reproduce in Portugal what is happening in Brazil. There is a strong possibility of expanding the use of workflow software not only in SDB but to other Grupo Sonae enterprises.

Appendix: a short history of process management

We give here a brief account of the origins of some of the main ideas behind process automation. We have highlighted those words and phrases that refer to important constituents of BPM.

Humans have made and used tools since we first emerged as a species about 125,000 years ago. Flint and other rocks gave us the raw materials for axes, knives and other edge tools. Bronze and then iron slowly supplanted flint.

Machine logic

Machines are ancient, too. Among the earliest were the potter's wheel and the bow drill, both of which originated some 6,000 years ago. The lathe came along about the same time. All these needed human muscle power.

Self-managed machines that draw on other sources of power have also been in use for many years. The ancient Egyptians had a water clock, although one could debate if this counts as a machine. Water-powered trip-hammers were in use in China from at least the 3rd century CE. Mechanical clocks began to appear in the thirteenth century CE.

These machines used simple, *straight-line logic*. In other words, they carried out an unvarying sequence of activities, which their makers defined when constructing them. The machines did just that one thing until they ran out of fuel or raw material, or broke down. Even the celebrated Jacquard loom, invented in 1801, simply recycled its chain of punched cards until it was switched off. Seen as a forerunner of computers, this was the first widely used machine that was separately *programmable*.

The use of *conditional logic* is also historical in origin. Mechanical governors to regulate the speed of mills, grindstones and, later, steam engines go back to the fifteenth century. In these, the device's actions vary according to the state of something outside itself. Governors thus provide *feedback*. (The idea of feedback is old but the word itself dates only from 1920.)

The development of computers

The next step was to create a machine able to *feed its results back to itself*, in what is called 'recursive working'. Although he never managed to build it, in 1834 Charles Babbage designed an 'Analytical Engine' incorporating this principle. Babbage's main innovation was to take the results of one series of calculations and use them in the next series. He described this as 'the machine biting its own tail'. All modern digital computers do this but it took another hundred years for the idea to become practicable.

Alan Turing made the next fundamental advance. In a 1936 paper, he showed that any mathematical calculation could be made through a series of small binary operations. This was achieved through an *algorithm* or procedure, which is the basis of a computer program. Programs allow digital computers to *model* activities in the outside world.

Using computers to do this modelling happened quickly afterwards. The first fully functioning electronic digital computer in the world was Colossus, built for code breaking in 1943. Only eight years later, a British baking and catering company developed its own commercial computer. J. Lyons & Co. built the LEO (Lyons Electronic Office) machine in 1951. It used this to do the management accounts and payroll for to its chain of teashops.

Another job was to *schedule* deliveries to the shops. This allowed a later and faster reaction to changes in orders. As Frank Land, one of LEO's developers, puts it: 'In other words the response to market conditions became closer to *real time*' (the italics are ours).

In 1955, a LEO computer began running the payroll of the British subsidiary of the Ford motor company. This was probably the first instance of *outsourcing* computer work.

In the half-century since then, the increasing *digitization* of data has transformed the way we create, transform, move and store data. Without it, modern process management would be impossible. We nowadays routinely use computers for numerical data, scanned images and word processed documents, for example. Most public telephone networks are digital, as are cellular telephones. So, too, are many televisions, increasing numbers of radios, many cameras, compact discs, satellite services and, of course, the Internet.

Industrial processes

Developments in factory automation have run broadly parallel to these scientific and administrative advances. Improving processes has been a preoccupation in manufacturing since before the Industrial Revolution. In fifteenth-century Venice, the city arsenal had shift workers producing weapons round the clock. Venetian warships had *interchangeable* parts.

Sea warfare also led to the world's first steam-powered factory, the Block Mills at Portsmouth Dockyard. These began working in 1803. The mills housed machinery designed by Marc Brunel (father to Isambard) to make wooden blocks for ships' pulleys. This was the first instance of *automated* mass production.

Henry Ford famously took this idea to an extreme at his Highland Park car works. In 1913, he combined interchangeable parts with standard work and moving conveyors to create what he called *flow production*. (Fellow carmaker Ransom E. Olds had introduced the moving assembly line in 1901 but without conveyors.)

Ford separated decision-making from work, an approach also being advocated at that time by Frederick Winslow Taylor. Taylor stressed the division of work into repetitive tasks and introduced the idea of *measuring and costing* work activities.

In 1947, Delmar Harder set up the Automation Department at Ford and popularized the word, '*automation*'. Within two years, the company was building factories able to make fully automated methods. These included mechanized handling and transfer devices, which were under the guidance of panels of relays, like those found in telephone switchboards. This was a forerunner of *computer-controlled production*.

At about the same time, Toyota Motors in Japan created the Toyota Production System. This changed the emphasis from individual machines and how they were used. Instead, it concentrated on the *end-to-end flow* of products. It measured *outputs*, not inputs or throughputs, and *reintegrated decision-making and work*. The Toyota system permits rapid changes to products while minimizing inventory and maintaining efficiency, a technique often referred to as *agile* manufacturing.

Office processes

Back in the white-collar world, computer companies were making products to digitize paper documents and manage the *routing* of the resulting files. Called DIP, this began in the 1980s. So, too, did producing separate *workflow management* software, to automate any kind of office procedure.

Pioneers of DIP tools included companies such as Plexus and Filenet, which originally sold their products as complete, closely integrated systems comprising hardware, software and networking. The Plexus software was incorporated into their product ranges ('OEMed') by NCR and Hewlett-Packard, among others. Olivetti was one of the main European resellers of Filenet products.

Specialized workflow software was rare at first. Two early suppliers came from the British Isles. Workhorse, of Dublin, developed some well-regarded software, also called Workhorse. AT&T used it in its 1990 office automation product, Rhapsody. Aldus, then a leading supplier of desktop publishing software, bought Workhorse in 1992, itself being taken over by Adobe Systems in 1994.

The other British workflow pioneer was the London-based company, Financial and Corporate Modelling Consultants (FCMC), founded in 1980. It developed procedure processing software – the forerunner of workflow – in 1985. FCMC followed this in with the first version of the Staffware product. It sold this direct and via IBM, ICL and several other suppliers. FCMC became Staffware PLC in 1993 and, in 2004, part of TIBCO.

In 1990, Geary Rummler and Alan Brache published a book called *Improving Performance: Managing the White Space on the Organization Chart*. This has some claim to have ignited general interest in process improvement. It introduced the term 'white space' to describe the *interfaces* between organizational functions.

That same year, Michael Hammer wrote an article in *Harvard Business Review* called 'Reengineering Work, Don't Automate, Obliterate'. He followed it three years later with a jointly written book on the same theme, called *Reengineering the Corporation*. These works popularized the *cross-functional design* of business processes and the idea of appointing *process managers*. Less happily, as we discussed in chapter 2, it fostered the notion that profound change should be carried out destructively.

In the sense we use it this book, the term 'Business Process Management' seems first to have appeared in print in 1993, in an article by Frank Leymann and Wolfgang Altenhuber. Both men were working for IBM, on its FlowMark workflow software. It was not until 2000 that the expression entered the vocabulary of most software sellers and consultants.

As you can see, much of what is wrapped up in the notion of BPM is well established, if not old. This is true of nearly all computer software and the way it is used. There are, however, also some important new ingredients in any comprehensive BPM software, which we discussed in the body of the book.

Glossary of BPM technical terms

We have included in this glossary the computer and other terms you are most likely to encounter in reading about and engaging in BPM. If you seek further or other definitions of technical words or initialisms, two excellent online resources are the *What Is* Web site, at http://www.whatis.com, and the *Free On-line Dictionary of Computing*, at http://foldoc.doc.ic.ac.uk.

AIIM	Association for Information and Image Management. American trade group of image and document management suppliers.
Algorithm	Originally, the mathematical concept of following a written process to achieve a goal. These days applied also to a set of computer instructions to perform a particular task. Named after the twelfth-century Iraqi mathematician, Muhammad ibn Musa al-Khwarizmi.
API	Application program(ming) interface. A set of standardized methods and formats to permit application programs to use the data and services of other software.
App	Slang abbreviation of 'application program'.
Application	Shorthand for an application program, the software that converts business requirements into computer commands. (Properly speaking, the application is the job the application software is there to do.)
Asynchronous	Delayed or allowed to be. Store-and-forward-working is asynchronous. Real-time, by contrast, is synchronous.
ATM	1. Automatic teller machine. The 'hole in the wall' bank terminal. First put into public use by Barclays Bank, in Britain, in 1969.
	2. Asynchronous Transfer Mode. A method for the high-speed transmission of packets of digital data.
Backbone	A bus (or hub or trunk) for connecting dissimilar communication systems or computer programs.
BAM	Business activity monitoring. Giving people a real-time view of active business processes and their outcomes.
Best of breed/best in class	A specialized or single-purpose product, supposedly superior as a result of being so.
Black box	Figuratively speaking, a system or part of one whose constituents you do not know or need to know. All you need knowledge of is its inputs and outputs.

Browser	The client software in a World Wide Web system that renders (transliterates) an HTML- or XML-coded document into something readable by humans.
Bullwhip effect	Descriptive of the large oscillations in the output of a system arising from comparatively small delays or variations in inputs. Usually applied to the amplified effects upstream caused by minor changes in demand in a supply chain.
Buy side	Systems to enable and manage online purchasing that are owned by the purchasing organization. They drive its e-commerce site if it has one.
Call-centre	An arrangement of several telephone lines and fewer people ('agents'), taking or making calls in rotation, such as for telesales, product support or customer enquiries.
Client-server	Cooperative processing between parts of an application program that are distributed over a network. Typically, the client component interacts with the user, while the server component deals with the data.
CMM	Capability Maturity Model. A stereotyped way of assessing the sophistication and formality of system design projects. Now being applied to business processes and organizational structures.
Cookie	A small file sent by a Web server to reside on a browser. It typically records customizing information that would otherwise have to be re-entered by the user.
COTS	Commercial off-the-shelf. Ready-made, packaged software.
Coupling	Describes the relationship between systems or sub-systems. Close (or tight) coupling means that changes in one directly and quickly affect the other(s). The elements are thus highly dependent on one another. Loose coupling offers comparative independence, allowing choice and incomplete communication between elements. Delay is sometimes the price of this. Soldering is close coupling, connecting by plug and socket is loose coupling. The components of a system architecture should be loosely coupled.
CRM	Customer relationship management. The use of computers to help in dealing with customers. *See* SFA.
CTI	Computer–telephony integration. The linking of telephone and computer systems to permit the control of voice storage and switching by the computer. Also the reverse, permitting voice control of computing activities.
Customer-facing	Anything requiring or involving contact with customers. Sometimes contrasted with the equally pretentious 'product-facing'.
Data mining	Extracting information from data. By examining statistics and trends, forming inferences about the meaning of the data held in company computers.
Data warehouse	Central store for all or most of an organization's business data. Known to IBM and its customers as an 'information warehouse'. Small versions, for special or local use, are called data marts.
Database	A set of logically connected files that have a common access point.

DBMS	Database management system. Software for controlling the input, output and upkeep of a computer database.
Decision tree	A logical table or diagram that sets out a sequence of (usually) binary decisions (that, is, requiring 'yes' or 'no' at each decision point). Trinary ('yes', 'no' or 'something else') or more varied decision choices are possible but make the tree harder to describe and obey.
Deterministic	Descriptive of something, such as a system, that exactly follows a defined path or sequence.
DIP	Document image processing. The creation and management of scanned images of paper documents. Usually linked to transaction or 'line of business' application programs, such as for insurance claims processing, and to workflow systems. *See* LOB.
Domain	An area on a system or network that can be managed as a single entity
EAI	Enterprise applications integration. Joining together an organization's main computer systems so that they can share and exchange data efficiently.
EDI	Electronic data interchange. The inter-company exchange of business data, carried direct from computer to computer. Typically deals with trade documentation such as purchase orders, invoices and shipping forms.
EDM	Electronic document management. The use of computers to create and manage sets of unstructured electronic documents. *See* unstructured.
E-forms	Electronic forms. *See* Electronic forms.
EIS	Executive information system. Originally intended to provide senior managers with high-level extracts and analyses of business data. Has since been democratized.
Electronic forms	Software that extracts data from corporate data stores and presents it in an easy-to-read style, often imitating a paper form.
Electronic mail	A means of conveying messages and associated files between human beings. Runs either on its own or as part of a corporate or external messaging service. Works in an asynchronous or 'store-and-forward' mode.
Electronic messaging	A store-and-forward network service for conveying data from one entity, such as a person or a program, to another. Embraces email but often, unhelpfully, used as a synonym for it.
Encryption	The encoding or scrambling of data, in a message for instance, so that only the sender and the intended recipient(s) can decipher ('decrypt') it.
Environment	The combination of operating system and related software on which something will run. *See* platform.
ERP	Enterprise resource planning. A class of application software that, originally, provided central control of and a unified systems environment for the computer systems in manufacturing companies. Has extended into most other sectors.
Extranet	A corporate Intranet to which outside persons and organizations are admitted.

Extreme programming	An interactive, evolutionary and team-based programming approach designed to produce better software more cheaply. Despite the name, does not require its practitioners to wear baggy trousers and sunglasses.
Filter	In email clients, user-defined set of parameters to dictate the destination of incoming messages, such as to particular directories or folders.
Granularity	Physical or figurative smallness of component parts. High granularity allows small increments.
GRC	Governance, risk and compliance. Usually refers to software that helps organizations improve the first, reduce the second and ensure the third. Mainly found in closely regulated industries such as finance and pharmaceuticals but spreading everywhere. Sarbanes–Oxley and Basel II are two examples.
Groupware	Software that helps people collaborate.
Heuristics	A way to write software that learns as it goes. These programs work to rules of thumb or guidelines, as opposed to precise and unchanging procedures. The word comes from the Greek for 'to find'.
Host	The computer on which server software runs. Also a verb describing that relationship.
HTML	HyperText Markup Language. The standard method for defining electronic pages for the World Wide Web. A dialect (purists would say an application) of SGML. *See* also XML.
HTTP	HyperText Transfer Protocol. The standard command method for summoning material or actions over the World Wide Web.
Infrastructure	The underlying foundation, whether literal or figurative, for an activity or system. The Internet is the infrastructure of the World Wide Web, for instance.
Instance	A concrete or tangible example of a process. You are an instance of the human life cycle, for example.
Internet	A specific federated network of computer networks that use the TCP/IP protocol. Consists of various national backbone nets and multitudinous regional, private and university networks around the world.
internet (small 'i')	Generic term, meaning any collection of networks that can function as a single, large virtual network.
Intranet	Internal (or virtual private) network based on Internet standards, especially those pertaining to the World Wide Web. *See* Extranet.
ISP	Internet service provider. A business that retails access to the Internet. Some only 'sell bandwidth'; others provide associated services.
Iteration	A repetition or running of a process or cycle.
JAD	Joint application development. The idea that system development is both faster and more effective if all the people involved work on the project together and at the same time.
Java	Programming language from Sun Microsystems. Extensively used on the World Wide Web and in intranets.
LAN	Local-area network. A system for linking computers that are physically close to each other, typically within the same building or site.

Linux	(Pronounced 'linnucks'.) Non-proprietary version of the Unix operating system, originally created by Linus Torvalds while a student in Helsinki.
LOB	Line of business. IBM-favoured term for what an organization does. LOB systems are computers running software specific to an organization's main activities.
Loosely-coupled	*See* coupling.
Message queuing	A kind of middleware that takes a store-and-forward approach, putting client requests into a closely managed queue and passing them to servers as they become free. *See* publish and subscribe.
Metadata	Data about data. A description or index of other data.
Middleware	Software that mediates between client and server processes in distributed systems. Provides central services over the network and a unified set of application programming interfaces.
Minitel	French public videotex service, begun in 1984 as an online telephone directory service.
Monolithic	Descriptive of software that the buyer cannot separate into its component parts or replace piecemeal. From the architectural term for something made from a single block of stone.
MRP	1. Materials requirements planning. Computer software to aid manufacturers in production planning.
	2. Manufacturing resource planning. Extension of the above idea to apply to other variables, such as employee numbers. The precursor to ERP. Also known as MRP II.
Net, The	or The Net. The Internet.
Object-oriented	Software written to use objects rather than data alone. Objects are packages incorporating a small amount of data with the methods needed to read or change it. Objects can communicate and be communicated with.
OLAP	Online analytical processing. Son of EIS. Provides multi-dimensional manipulation of extracts from business data. Has popularized the term 'metadata' – data about data.
OLTP	Online transaction processing. The high-speed handling of LOB application data.
OMG	Object Management Group. A cross-industry standards body concerned with interoperability between object-oriented software and systems.
Orchestration	In Web services (see below), managing and coordinating the assembly of component services to create a complete business process.
PC	Personal computer. A general-purpose machine intended for use by one person. Contrary to popular perception, an Apple Macintosh is a PC.
PDA	Personal digital assistant. A personal computer small enough to hold comfortably in one hand.
PDF	Portable Document Format. File definition format used by Acrobat in its Adobe Acrobat electronic paper product.
Peer-to-peer	A way of managing computer networks such that there is no central server, each point on the network being able to import, store and export data. Popularized by MP3-sharing systems such as Napster.

Person-to-person	Direct and, by implication, real-time computer-aided communication between people.
Platform	1. Anything, such as an operating system, microprocessor or network, on which the thing being talked about will work. *See* environment.
	2. The type of computer or operating system upon which an application program will run.
Portal	A doorway, both in ordinary life and on the Web. On the latter it refers to a site that is or purports to be a major way station for users, either on the Web or on an intranet.
Procedure	A sequence of actions and events that conforms to a set of explicit instructions.
Process	A sequence of actions and events that, consciously designed or not, aims to achieve a purpose.
Protocol	A description of the messages to be exchanged and rules to be followed when two or more computer systems exchange information.
Publish and subscribe	Middleware that, as its name suggests, lets servers 'publish' the existence of processes or other software entities, to be requested ('subscribed to') by clients. *See* message queuing.
Purchasing card	Company credit card, issued in multiples for use by an organization. Allows authorized employees to make cashless and, often, paperless purchases from suppliers.
RAD	1. Rapid application development. The use of a modelling language, normally visual, to produce applications software. Usually involves prototyping.
	2. Role activity diagram. A way of recording work activities in systems analysis.
Real-time	Happening soon or quickly; with little delay or lag.
Replication	The reciprocal copying of database records between distant machines. Is the basis of Lotus Notes' data distribution method.
Requisite variety	The idea that, if a system is to deal adequately with its environment or another system, it must be able to match the number of states it possesses. Those states are typically represented by different information types or levels.
RFID	Radio frequency identification device. Very small radio transmitters affixed or inserted into something, typically a manufactured product, to allow it to be tracked electronically.
Routing	The ability of software to send a file or message to a preselected list of recipients. These may be people, roles or machines.
RPC	Remote procedure call. Software tools and routines to allow application programs to interwork directly with other computing entities over a network.
RSS	Really simple syndication. A way to broadcast information from Web sites.
Rules	A way of applying variable logic to application software. Rules range from those to file personal email to those behind complete, organization-wide business processes. Rules can be predefined or change as software learns. *See* heuristics.

Rules engine	Software that subjects input data to a sequence of tests and decisions – a rule. The output data typically includes instructions to an application program.
SCM	1. Supply chain management. Application software for managing the flow of commercial data, especially that relating to purchasing and supply, between trading entities or companies.
	2. Software configuration management. Controlling and keeping track of software development.
SCOR	Supply Chain Operations Reference. A template of the ingredients of a supply chain, devised and maintained by the Supply Chain Council.
Script	A program or set of instructions that is run by another program rather than directly by the computer's processor. Easier to write than full programs but are less efficient.
SFA	Sales force automation. Computer-aided selling. Ranges from contact management software on a laptop PC to company-wide systems that permit team coordination, role swapping and visit control.
SGML	Standard Generalized Markup Language. An international standard for annotating text for electronic publishing. *See* HTML, XML.
Silo	Figurative term for a system that works only within its own organizational unit (by analogy with a grain silo).
Simulation	Building and running a dynamic model of a process, to help people understand it better and improve it.
SMS	Short Message Service. Email system for digital cellular telephone services, each message occupying a maximum of 160 characters.
SOA	Service Oriented Architecture. Splitting the design of the data architecture of a system from the processes needed to run it. In theory, having the two layers loosely coupled makes it easier to make and manage them. *See* coupling.
SOAP	Simple Object Access Protocol. A method for constructing XML messages to be exchanged within Web services. *See* Web services.
SQL	Structured Query Language. A loosely standardized method, originated by IBM, for formulating requests to a relational database.
Step	A stage or event in a business process that calls for a human or computer response or interaction of some kind. Also known as an instruction, node, operation, process element, task, and work element.
STP	Straight-through processing. Automatic handling of data and messages except where needed for policy reasons. The banking sector's term for end-to-end working.
Swimlane	The strip on a flowchart that represents a person, group, machine or role or group. Occurs in functional (or cross-functional) and deployment charts. So called because it supposedly reminds one of the lanes in a swimming pool.
System	1. An entity that maintains its existence through the mutual interaction of its parts.
	2. A shorthand term for a computer system.
TCP/IP	Transmission Control Protocol/Interconnect Protocol. A client-server networking method, used on the Internet and for Unix-based systems.

Thin client	A computing architecture based upon holding the minimum amount of data at the client. Computing logic and the rest of the data are held on the server. A Web browser is a thin client. *See* client-server.
TPM	Transaction processing monitor. TPMs control the flows into, and update, multiple database servers, often from large numbers of clients. They are typically found in high-throughput central systems.
UML	Unified Modelling Language. A graphical language for describing and defining processes.
Unstructured	A computerist's description of data in a document that a computer cannot make sense of.
VPN	Virtual private network. A partitioned-off, and relatively secure, path over the Internet or public telephone network, for the use of known participants. Contrasts with a leased line arrangement.
W3C	World Wide Web Consortium. Body charged with developing and enforcing standards for the Web. Sir Tim Berners-Lee is head prefect. *See* WWW.
WAN	Wide-area network. Uses public and private circuits to create an enterprise network, spanning a domain greater than a typical local area network can deal with.
WAP	Wireless Application Protocol. A set of standards for linking wireless devices, such as cellular telephones, to the Internet.
Web browser	*See* browser.
Web page	A collection of linked computer files and screen layouts that appears in a Web browser as a single, continuous image.
Web server	The machine on which reside the Web pages and associated data to service a company's Web site. Sometimes called the host.
Web services	A way of creating on-demand business processes that run across connections made over the World Wide Web or an Intranet. Involves the use of orchestration, SOAP and XML. (We show the term with capitals to differentiate it from any other kind of service offered over the Web.)
Web site	A set of Web pages, typically held on a single server, and indexed and treated as a unity. Has its own domain name and number.
Web, The	The World Wide Web. *See* WWW.
WfMC	Workflow Management Coalition. The world's workflow software makers in conclave. Has produced a set of APIs to allow interworking between dissimilar products.
Workflow software	Models human work processes on a computer, enforcing compliance with established sequences and procedures.
World Wide Web	WWW. More familiarly known as 'The Web'. A graphical information retrieval system that operates over the Internet. Its standards and methods are also usable for internal networks (intranets) and private areas on the Internet (extranets).
XML	eXtensible Markup Language. A simplified version of SGML that is more powerful, robust and adaptable than its sibling, HTML.

Suggested reading

Here are some books and Web sites from which we think you would gain value. They are grouped into five broad categories – systems thinking, organizational behaviour, the business case, business process improvement and further resources.

This list is neither complete nor unbiased. Also, several of the publications are quite old; they are publications and places we keep going to for ideas, inspiration and encouragement. These do not date. Some of the sources we suggest are for broadening the mind; others are narrower in scope.

For further suggestions and other material, we recommend that you:

- Call on your local library. This could be your company library, if you have one, that for your professional institute or, in large towns and cities, the public library. Most librarians are keen to help and can often put you in touch with people who can help you further. You may need to go to a library to get hold of some of the books we mention.
- Go to the Web sites of the main online booksellers and see what is popular and what readers recommend (not always the same thing). Take readers' reviews with a pinch of salt; they are not always what they seem.
- Visit Yogesh Malhotra's BRINT Web site, at http://www.brint.com. He and his colleagues have created a wide-ranging index of business knowledge that links to a huge amount of source material. Use of it is free.
- Consult the Wikipedia, at http://en.wikipedia.org/wiki/Main_Page. This is an online encyclopaedia compiled and updated by its readers, and is also free. Despite what you might expect, the quality of articles is high. The range is broad and the linking between entries is excellent.
- For the latest on-line news and comment, use an RSS aggregator. Not as fearsome as it sounds, this is software that collects material as it appears on those Web sites you wish to keep an eye on. Depending on which software you choose, it presents the results in a separate window on your screen, in your Web browser or in your email access software. Many RSS readers are free.

Systems thinking

1. Peter Checkland, *Systems Thinking, Systems Practice*, Chichester: John Wiley, 1999. The primary and original text on 'soft systems' thinking – the kind that's harder to do. Readably

combines theoretical explorations with practical examples and advice. Is the antithesis of the painting-by-numbers approach found in many other methods. First published in 1981; now available in paperback, with a 'thirty-year perspective' included.

2. Stafford Beer, *The Brain of the Firm: The Managerial Cybernetics of Organizations*, Chichester: John Wiley, 1994. This is the first of the three books in which Beer expounds his model of the organization as a living being. This is not an overnight read, and contains some (skippable) mathematics, but is packed with insight.

3. Sir Geoffrey Vickers, *Human Systems are Different*, London: Harper & Row, 1984. Vickers was, at various times, a soldier (winning the Victoria Cross), a lawyer and a chairman of the then National Coal Board (NCB) in Britain. His understanding of systems behaviour reflected this varied and non-technical background. It resulted in some of the best writing there has been on the subject. Like all his books, *Human Systems are Different* deals with the human, organizational and political spheres. We list it as a contrast to the sometimes over-abstract offerings that other authors put before the reader.

Organizational behaviour

1. Richard Ritti and Steven Levy, *The Ropes to Skip and the Ropes to Know: Studies in Organizational Behavior*, Chichester: John Wiley, 2002 (sixth edition). Provides vivid and memorable insight into organizational behaviour through a series of fictitious stories, a formula later copied by Goldratt and others. Although based mainly on American manufacturing industry, it points out as clearly as anything more modern the reefs on which an BPM strategy might founder.

2. Charles Hampden-Turner and Fons Trompenaars, *The Seven Cultures of Capitalism: Value Systems for Creating Wealth in the United States, Britain, Japan, Germany, France, Sweden and the Netherlands*, London: Judy Piatkus, 1994. A practical, entertaining and thought-provoking journey around the world's main industrial cultures. Combines the results of extensive surveys with historical sketches, case studies and anecdotes.

3. Gareth Morgan, *Images of Organization*, Beverly Hills, CA: Sage, 1986. A penetrating analysis of the prevailing mental models of organizations, giving the strengths and weakness of each. Finishes with a less convincing synthesis of the nine metaphors discussed. This scarcely detracts from the mind-opening material that precedes it.

4. Ricardo Semler, *Maverick!*, New York: Random House, 2001 (first published in 1993). Describes how Semler turned upside down the conventional rules in his family's business and succeeded despite it. An ego trip but exciting and refreshing for all that. Semler has written a sequel, *Seven-Day Weekend* (London: Century, 2003) that vindicates the approach. It also deals with matters not covered in the earlier book, such as home working and electronic mail.

5. Charles Handy, *Understanding Organizations*, 4th edn., Harmondsworth: Penguin Books, 1992. A classic. Handy covers all the important topics in a balanced but shrewd fashion, with numerous examples in every chapter. The best general introduction to organizational matters available but could do with updating. A chapter on the effects of the Internet would be valuable.

The business case

1. Paul Strassmann, *The Business Value of Computers:*, The Information Economics Press, 1990. A detailed investigation of various accounting methods for IT investments, leading to an explanation of Strassmann's notion of management value added as a robust measure. A big book, only for the committed reader. Strassman has published several other useful books on the theme of the value of IT. He has a Web site with some of his articles on it.
2. Barbara Farbey, Frank Land and David Targett, *How to Assess Your IT Investment: A Study of Methods and Practice*, London: Butterworth Heinemann, 1993. A practical handbook, published in association with the magazine *Management Today*. The authors cover most of the problems in succinct fashion, including grasping the nettle of qualitative benefits.

Business process improvement

1. Robert Townsend. *Up the Organization: How to Stop the Company Stifling People and Strangling Profits*, London: Coronet Books, 1970. Down-to-earth advice wittily presented by the man who resurrected Avis. It has dated only in some of the detail. Townsend has no time for flim-flammery or for activities, departments or ideas that don't earn their keep. Speaks more common sense on reengineering – but before the word was invented – than any of his successors, and in far fewer words.
2. Paul Harmon, *Business Process Change: A Manager's Guide to Improving, Redesigning, and Automating Processes*, San Francisco: Morgan Kauffman, 2003. Mainly an engineering textbook but the best all-in-one guide to BPM trends and technologies we know of. Gives copious and well-illustrated detail on all the popular methods and taxonomies. You need to look elsewhere for coverage of divergent viewpoints, different organization models, cultural forces and other aspects of human behaviour.
3. *Business Process Trends* (see http://www.bptrends.com/index.cfm). Managed by Paul Harmon, this is a free Web site on BPM. It is strongly technocentric, like his book (see above), giving little coverage to broader system, human and organizational considerations. There is a monthly newsletter, email updates and news.
4. Gary Rummler and Alan Brache, *Improving Performance: Managing the White Space on the Organization Chart*, San Francisco: Jossey-Bass 1990. This is a favourite with many people as a practical guide. The authors present their arguments and guidance clearly and in a logical sequence. Although obviously not covering the possibilities that modern BPM software offers, it is still relevant. Its main defect is that it starts too low in the Meadows list of places to intervene in a system (see chapter 8).

Further resources

Here are links to some Web sites you may find useful. They include research groups, trade and professional organizations and academic study groups.

AIIM (Association for Information and Image Management) – http://www.aiim.org

American Society for Cybernetics – http://www.asc-cybernetics.org

BFMA (Business Forms Management Association) – http://www.bfma.org

BPMG (Business Process Management Group) – http://www.bpmg.org

BPMI (Business Process Management Initiative) – http://www.bpmi.org

BPR and Change Management Learning Center – http://www.prosci.com

Business Activity Monitoring.com – http://www.businessactivitymonitoring.com

Enterprise Workflow National Project (UK) – http://www.workflownp.org.uk

Howe School of Technology Management – Research in Process Management and Workflow Automation – http://attila.stevens.edu/workflow

ISO (International Organization for Standardization – see its ISO9000 and ISO14000 pages) – http://www.iso.org

OASIS (Organization for the Advancement of Structured Information Standards) – http://www.oasis-open.org

OMG (Object Management Group) – http://www.omg.org

SIGPAM (Special Interest Group on Process Automation and Management) – http://www.sigpam.org

SysWeb (Open University Systems Group) – http://systems.open.ac.uk

WARIA (Workflow and Reengineering International Association) – http://www.waria.com

WfMC (Workflow Management Coalition) – http://www.wfmc.org

References

Anonymous. *Industrial Distribution*, 1 December 1998, see http://www.manufacturing.net/ind/article/CA115716.html

Aponovich, D., 'Era of the E-Business Ecosystem', *Datamation*, 23 April 2002, see http://itmanagement.earthweb.com/ecom/article.php/1014411

Argyris, C., *Overcoming Organizational Defenses*, Englewood Cliffs, NJ: Prentice-Hall, 1990

Ashby, R., *An Introduction to Cybernetics*, London: Chapman & Hale

Beer, S., *The Brain of the Firm: The Managerial Cybernetics of Organizations*, Chichester: John Wiley, 1994

Bleed, R., *A Learning Organization*, see http://www.dist.maricopa.edu/users/bleed/learn3.htm

Capra, F., 'The Role of Physics in the Current Change in Paradigms', in R. F. Kitchener (ed.), *The World View of Contemporary Physics: Does it Need a New Metaphysics?*, New York: State University of New York Press, 1988, pp. 144–55

Carter, R., J. Martin, W. Mayblin and M. Munday, *Systems, Management and Change: A Graphic Guide*, London: Paul Chapman, 1984

Davenport, T. H., 'The Fad that Forgot People', *Fast company*, 01, November 1995, p. 70

De Geus, A., *The Living Company: Growth, Learning and Longevity in Business*, London: Nicholas Brealey, 1997

Diderot, D., *Oeuvres romanesques*, Paris: Classiques Garnier, 1962

Drucker, P. F., *Landmarks of Tomorrow*, New York: Harper, 1959

Exigen Group, *Spotting the Symptoms of Corporate Cholesterol: A Guide to Identifying Process Inefficiencies in the Insurance Sector*, May 2004, see http://www.exigengroup.com/cholesterol

Greiner, L., 'Evolution and Revolution as Organizations Grow', *Harvard Business Review*, July–August 1972, pp. 37–46

Hammer, M., 'Reengineering Work, Don't Automate, Obliterate', *Harvard Business Review*, July–August 1990, pp. 104–12

Hammer, M. and J. Champy, *Reengineering the Corporation: A Manifesto for Business Revolution*, London: Nicholas Brealey, 1993

Handy, C., *Understanding Organizations*, Harmondsworth: Penguin Books, 1985 (4th edn., 1993)

Hauser-Kastenberg, G., W. E. Kastenberg and D. Norris, 'Towards Emergent Ethical Action and the Culture of Engineering', *Science and Engineering Ethics* 9(3), 2003

Hosier, J. (ed.), *The Handbook of IBM Terminology*, Newbury: Xephon, Spring 1995

IBM, *Patterns for e-business* Web site, see http://www-106.ibm.com/developerworks/patterns/select-pattern.html

Janecek, R., *Nothing Impersonal*, 19 October 2003, see http://radovanjanecek.net/blog/archives/000026.html

Kent, W., *Data and Reality*, New York: North-Holland 1978

Keynes, J. M., *The General Theory of Employment, Interest and Money*, London: Macmillan, 1936

Kohn, A., *Punished by Rewards: The Trouble with Gold Stars, Incentive Plans, A's, Praise, and Other Bribes*, Boston, MA: Houghton-Mifflin, 1999

Land, F., 'LEO, the First Business Computer: A Personal Experience', in R. L. Glass (ed.), *In the Beginning: Recollections of Software Pioneers*, New York: IEEE Computer Society Press, 1998, pp. 134–53

Levitt, T., *The Marketing Imagination*, New York: Free Press, 1986

Leymann, F. and W. Altenhuber, 'Managing Business Processes as an Information Resource', *IBM Systems Journal*, 33(2), 1994, see http://www.research.ibm.com/journal/sj/332/leymann.html

McGregor, D., *The Human Side of Enterprise*, New York: McGraw-Hill, 1960

McGuffog, T., *It's One-Part 'e' and Nine-Parts 'Business'*, Presentation to UK Council for Electronic Business, 20 February 2003, see http://www.ukceb.org.uk/cpRoot/306/6/Tom per cent20McGuffog per cent20presentation.pdf

Maritain, J., *The Peasant of the Garonne: An Old Layman Questions Himself About the Present Time*, TX: Austin, Holt, Rinehart & Winston, 1968

Mayo, E., *The Human Problems of an Industrial Civilization*, New York: Macmillan, 1933

Meadows, D. H., 'Places to Intervene in a System', The Sustainability Institute, Hartland, VT, 1999, see http://www.sustainer.org/pubs/Leverage_Points.pdf

Meadows, D. H., 'Dancing with Systems', The Sustainability Institute, Hartland, VT, undated, see http://www.sustainer.org/pubs/Dancing.html

Melymuka, K., 'Innovation Democracy', *Computerworld*, 16 February 2004, see http://www.computerworld.com/managementtopics/management/story/0,10801,90207,00.html

Merriam-Webster, Inc., *Collegiate Dictionary* online version, see http://www.m-w.com/dictionary.htm

Meyer, G., 'Jet-Fueled', *Optimize*, February 2003, see http://www.optimizemag.com/issue/016/leadership.htm

Micklethwait, J. and A. Wooldridge, *The Witch Doctors: What the Management Gurus are Saying, Why it Matters and How to Make Sense of it*, London: Mandarin Paperbacks, 1997

The MIT Forum for Supply Chain Innovation, *The MIT Beer Game*, see http://beergame.mit.edu/gameapplet.asp

Morgan, G., *Images of Organization*, Beverly Hills, CA, Sage, 1986

Ouchi, W., *Theory Z: How American Business Can Meet the Japanese Challenge*, New York, Basic Books

Oxford University Press, *Oxford English Dictionary* (2nd edition) on CD-ROM, version 3.0, 2002

Pascale, R. T., M. Millemann and L. Gioja, *Surfing the Edge of Chaos: The Laws of Nature and the New Laws of Business*, New York: Three Rivers Press, 2001

Peters, T., *The Tom Peters Seminar: Crazy Times Call for Crazy Organizations*, London: Macmillan, 1994

Roethlisberger, F. J. and W. J. Dickson, *Management and the Worker*, Cambridge, MA: Harvard University Press, 1939

Rummler, G. A. and A. P. Brache, *Improving Performance: Managing the White Space on the Organization Chart*, San Francisco: Jossey-Bass, 1990

Schmenner, R., 'Some Measures of Concern', *Business Review Weekly*

Semler, R., *Maverick!*, London: Century, 1993

Semler, R., *The Seven-Day Weekend*, London: Century, 2003

Sinur, J., *Business Process Management's Next Big Adventure*, Presentation at the Gartner IT Symposium/ITxpo, Orlando, Florida, 6–11 October 2002

Smith, H. and P. Fingar, 'BPM's Third Wave: Build to Adapt, Not Just to Last', *ebizQ*, 20 January 2003

Standage, T., *The Victorian Internet: The Remarkable Story of the Telegraph and the Nineteenth Century's Online Pioneers*, London: Weidenfeld & Nicolson, 1998

Strassmann, P., *In formation pay off*, New York: Free Press, 1985

Tannenhaum, R. and W. H. Schunot, 'How to Choose a Leadership Pattern', *Harvard Business Review*, March–April 1958, pp. 95–101

Tapscott, D. and A. Caston, *Paradigm Shift: The New Promise of Information Technology*, New York: McGraw-Hill, 1993

Taylor, F. W., *The Principles of Scientific Management*, New York: Harper, 1911

Trompenaars, F., *Riding the Waves of Culture: Understanding Cultural Diversity in Business*, London: Nicholas Brealey, 1993

Turing, A. M., 'On Computable Numbers, with an Application to the *Entscheidungsproblem*', Proceedings of the London Mathematical Society, 2 (42), 1936

Vickers, G., *Human Systems are Different*, London: Harper & Row, 1984

Vitria Technology, Inc., Case study: 'Availity Painlessly Delivers Collaborative Healthcare', 2003, see http://www.vitria.com/library/case_studies/vitria_casestudy_availity.pdf

Wiener, N., *Cybernetics, or Control and Communication in the Animal and the Machine*, New York: John Wiley, 1948

Womack, J. P., 'Lean Thinking: A Look Back and a Look Forward', Lean Enterprise Institute, undated, see http://www.lean.org/Community/Registered/Article.cfm?ArticleId=23

Youngblood, M. D., *Eating the Chocolate Elephant: Take Charge of Change through Total Process Management*, Richardson, TX: Micrografx, Inc., 1994

Index